Urban Mobilities in the Global South

Edited by
Tanu Priya Uteng and
Karen Lucas

Routledge
Taylor & Francis Group
LONDON AND NEW YORK

First published 2018
by Routledge
4 Park Square, Milton Park, Abingdon, Oxon OX14 4RN
605 Third Avenue, New York, NY 10017

Routledge is an imprint of the Taylor & Francis Group, an informa business

British Library Cataloguing-in-Publication Data
A catalogue record for this book is available from the British Library

Library of Congress Cataloging-in-Publication Data
A catalog record for this book has been requested

ISBN: 978-1-138-29171-3 (hbk)
ISBN: 978-1-315-26509-4 (ebk)

Typeset in Times New Roman
by Apex CoVantage, LLC

Contents

Figures

Tables

Contributors

Tanu Priya Uteng is Senior Researcher at the Institute of Transport Economics in Oslo, Norway. She has worked and published across a host of cross-cutting issues in the field of urban and transport planning for the past 14 years. A few of her areas of expertise include transport-related social exclusion, gendered mobilities, travel behaviour studies, transport modelling and econometric analyses. She is currently leading long-term strategic projects looking at topics such as bicycling, first-last mile connectivity, climate and travel behaviour, green-shift in the Nordic region, commuting and car-sharing. She is deeply interested in mobilities issues in the developing countries, and the singular focus with which *the wrong* planning solutions get copied across the world intrigues her. Her first edited book, *Gendered Mobilities*, was published in 2008, which was later used to inform a background paper for the 2012 World Development Report focusing on 'Gender and Mobility in the Developing World'.

Karen Lucas is Professor of Transport and Social Analysis at the Institute of Transport Studies, University of Leeds. She has 20 years of experience in social research in transport. She is a world-leading expert in the area of transport-related social exclusion, and she leads an international research network and regularly publishes on this topic. In 2015, she received the Edward L. Ullman Award by the Transport Geography Specialty Group of the Association of American Geographers for her significant contribution to transportation geography. Karen is a regular advisor to national governments in the UK. She has also advised local and national governments in Australia, South America, South Africa and New Zealand and is an active member of a number of advisory bodies.

Lake Sagaris is an internationally recognized expert on cycle-inclusive urban planning, civil society development and participatory planning theory and practice as they relate to urban-regional governance. An award-winning writer and editor, she began her working life in Chile in 1980 as a freelance journalist with the *London Times*, *Toronto Globe and Mail*, *Miami Herald* and other media. Her current work uses participatory action research methods and community-government partnerships in Santiago and Temuco (Chile) to apply an intermodal approach based on 'ecologies of modes and actors' to transition toward more sustainable transport. She focuses on resilience, social justice and inclusion, particularly gender, safety and security issues. These experiences have led to awards in Chile and abroad, participation in a UN Expert Group meeting on sustainable transport, and presentations in Latin America, Europe, Canada, Taiwan, the United States and India.

Noelani Eidse is a PhD candidate in the Department of Geography at McGill University, Montreal, Canada. She explores how underlying relationships between urbanisation, public space livelihoods and governance strategies are framed by subtle negotiations and everyday politics. Her doctoral research, funded by the Social Science and Humanities Research Council (SSHRC) and the International Development Research Centre (IDRC), investigates the daily life and survival of informal vendors working on the edge of legality. Noelani recently published her research on vendor mobilities in the Annals of the American Association of Geographers (2016), and has written on participatory methodologies in Area (2014).

Anvita Arora is an architect and transport planner with a PhD from the Civil Engineering Department of the Indian Institute of Technology (IIT), Delhi in 2007. She is the Managing Director and CEO of Innovative Transport Solutions (iTrans), an incubatee company of IIT, Delhi. The company has successfully delivered more than 40 applied research and planning projects in the past nine years under her leadership, to clients ranging from various city- and country-level authorities to funding agencies such as the UNEP, World Bank, Asian Development Bank, DFID and others. The primary focus of the work has been on supporting cities to become sustainable, inclusive and climate resilient.

Gina Porter is a professor in the Department of Anthropology, Durham University. Her research combines ethnographic approaches with a strong interest in spatial perspectives. Uneven power relationships and associated issues of exclusion are linking themes through her work. The majority of her field research has been in sub-Saharan Africa where her current research focuses principally on daily mobilities and the impact of mobile phones on young people's lives. Her latest publications include *Young People's Daily Mobilities in Sub-Saharan Africa: Moving Young Lives* (Palgrave Macmillan 2017).

Kate Hampshire is a medial anthropologist, with 20 years' research experience in sub-Saharan Africa. Her recent work includes research on trust across medicine supply chains in Ghana and Tanzania, and the 'informal' use of mobile phones for health care in Ghana, Malawi and South Africa.

Ariane De Lannoy heads up the Youth Research Stream at the Poverty and Inequality Initiative (PII) at the University of Cape Town. Since the start of her PhD research in 2004, her research work has had a consistent focus on youth and youth development in the complex context of post-apartheid South Africa, and has involved both quantitative and qualitative methods. Her work concentrates on teasing out the complex interrelationships among structural constraints confronting youth, their aspirations and their ability to take action. Ariane has consistently argued that it is imperative that policies directed at youth are based on a thorough understanding of this interplay, so that new, evidence-based interventions create possibilities to which young people are able and willing to respond.

Nwabisa Gunguluza was a researcher at the Children's Institute, University of Cape Town until mid-2015. Her research interests include ethnographic research, youth, literacy and linguistics. She has previously worked with Ariane De Lannoy on an ethnographic study with the Children's Institute to document the lives of ordinary youth in 'new' South Africa: *After Freedom: The Rise of the Post-Apartheid Generation in Democratic South Africa* (2014, Beacon Press, Boston). She has continued to collaborate with the Institute, including on the ESRC DFID-funded youth phones project led by Durham University.

Mac Mashiri is a spatial, land use and transportation planning and development consultant. He is a director at GTRD, a consulting firm based in Tshwane, South Africa. Before his current role, he worked for the Council for Scientific and Industrial Research (CSIR) in Tshwane for 14 years. Mac envisions his role as a consultant in terms of championing pro-poor built environment policies, strategies and plans that recognise and deliver on the vision and constituent needs and demands of Africa's people.

Andisiwe Bango has a BSc Honours degree from the University of Transkei and a Master of Environmental Management from the Free State University. He joined Walter Sisulu University's Department of Environmental Science in 2005. He has a particular interest in society and environmental relations, including water as an environmental resource and society as users. He has recently conducted research on community relations in protected areas and has just completed a degree in law from Walter Sisulu University.

Jennifer O'Brien is a lecturer and Director of Social Responsibility for the School of Environment, Education and Development, University of Manchester. She is a development geographer specialising in health, particularly maternal-child health, and has lived in rural Uganda for a number of years. Her current research focuses on how informal knowledge and participatory methodologies can provide solutions to long-standing development problems.

James Evans is a professor of geography at the School of Environment, Education and Development, University of Manchester. His research focuses on how cities learn to become more sustainable in the face of multiple challenges, and he currently leads projects on smart cities, informal mobility and resilience. He also leads the Smart and Sustainable Cities research theme as part of the Manchester Urban Institute.

Jacqueline M Klopp is an Associate Research Scholar at the Center for Sustainable Urban Development and teaches in the Sustainable Development Program at Columbia University. A political scientist by training, her work focuses on the political processes around land use, transportation, violence, displacement and planning in African cities. She is a founding member of DigitalMatatus.com and helped start the blogs CairofromBelow and nairobiplanninginnovations.com to provide more grounded and open urban information to citizens. Klopp received her B.A. from Harvard University where she studied physics, and her PhD in political science is from McGill University.

Clemence M Cavoli is a researcher at the Centre for Transport Studies at University College London. Her field of expertise and research interests are in environmental and sustainable mobility policymaking and planning, particularly at the urban level. She coordinates research across five Eastern European and Middle East cities (mainly capital cities) in the context of a large European project called CREATE that focuses on sustainable mobility across Europe and the Middle East. She also manages a project focusing on collective transport and sustainable mobility based in Maputo, Mozambique. She worked for the Department for Transport in the United Kingdom and the European Union Commission on urban transport policymaking.

Yusak Susilo is a professor at the Royal Institute of Technology, Stockholm. His main research interest lies in the intersection between transport and urban planning, transport policy, decision-making processes and behavioural interactions modelling. His research focuses on the variability of individuals' travel patterns and decision-making behaviours and the interaction of such patterns with changes in urban form, policies, environments and technology characteristics. He has been a principal investigator and co-investigator in various international and national projects, and has been actively exploring individuals' travel needs and behaviours and road safety issues in Southeast Asian countries for more than a decade.

Chengxi Liu is a researcher at the VTI Swedish National Road and Transport Research Institute. His main research interest lies in the individual behavioural modelling, data mining and transport demand estimation. He has been involved in several national and international projects. He utilises and develops various econometric and discrete choice modelling tools to study individual travel behaviour and travel demand. He has been studying travel behaviour responses to climate change, intra-household travel behaviour modelling, bicycle travel demand and last-mile logistic solutions.

Dirk Heinrichs leads the Department of Mobility and Urban Development at the DLR Institute of Transport Research and is Professor for Urban Development and Mobility at Technische Universität (TU) in Berlin. He is also the director of a joint master program on Urban Planning and Mobility located at TU Berlin and the University of Buenos Aires. Trained as urban and regional planner, his research focus is on the linkages among urbanization, household location choice and daily mobility as well as on innovative models of urban transport provision in Latin America, Asia and Africa. He currently leads a project which explores informal urban transport in Africa with support from the German Research Foundation.

Daniel Ehebrecht is a doctoral researcher at the Geography Department at Humboldt-Universität zu, Berlin. He studied social and economic geography and has worked as research assistant at the Institute for Geography and at the Institute for Migration Research and Intercultural Studies at Osnabrück University. His research interests relate to the nexus of urbanization, public transport and informality in Africa. In his PhD project he investigates governance structures of motorcycle-taxi services in Dar es Salaam, with a close look at actors, relations and social practices. His research is linked to a project which explores informal urban transport in Africa and is funded by the German Research Foundation.

Barbara Lenz is Director of the DLR Institute of Transport Research and Professor for Transport Geography at Humboldt Universität zu Berlin. One core topic in her research on transport demand and travel behaviour in the passenger and freight sectors is the implications and effects of technology use. These include extensive research on the interrelation of new information and communication technologies (ICT) and travel behaviour, the use of new platform-based mobility concepts and automated driving technology from a user perspective. Together with Dirk Heinrichs she currently leads a project which explores informal urban transport in Africa with support from the German Research Foundation.

Jesse Harber is a political economist with a specialisation in South African urban development and governance. He has worked on the Long-Term [Climate] Adaptation Scenarios for Human Settlements, the Cities Support Programme of the National Treasury and on studies of integrated ticketing and green transport. He is currently a researcher at the Gauteng City-Region Observatory, a partnership between the Gauteng Provincial Government, the Universities of the Witwatersrand and Johannesburg, and organized local government in Gauteng.

Daniel Oviedo is a researcher and lecturer in transport and development planning at University College London. Daniel is an experienced researcher and consultant in projects related to urban and interurban transport in cities of Latin America, Africa and India. He holds a PhD in Development Planning from University College London and a BSc and MSc in civil engineering from Universidad de los Andes, Colombia. Daniel specializes in the social, economic and spatial analysis of inequalities related to urban transport and policy evaluation in developing countries. His most recent project explores transport and well-being in urban Nigeria. Daniel has published several journal articles related to the links between transport, poverty, accessibility and social exclusion.

Caren Levy is a Professor of Transformative Planning of the Development Planning Unit, a research and postgraduate teaching department at University College London. She has more than 30 years of experience in research, consultancy, teaching and capacity building in urban development in Africa, Asia, the Middle East and Latin America, with a focus on urban planning, governance and community-led development. She has worked with governments, bi- and multi-lateral development agencies, universities and NGOs through innovatory approaches to planning methodology, organisational development, capacity building and learning related to the development of the built environment. Her most recent project explores transport and well-being in urban Nigeria. She has several publications on varied planning topics including 'deep distribution' and right to the city, specifically in the context of discussing development in cities of the Global South.

Julio D Dávila is Professor of Urban Policy and International Development and Director of the Development Planning Unit, a research and postgraduate teaching department at University College London. He has more than 25 years' experience in more than a dozen countries in Latin America, the Middle East, Africa and Asia. He worked at the International Institute for Environment and Development (London and Buenos Aires) and at the Colombian government's National Planning Department. Much of his research work focuses on the role of local government in progressive social and political transformation in developing countries; the governance dimensions of urban and peri-urban infrastructure, especially public transport, and water and sanitation; the intersection between planning and urban informality; and the linkages between rapid urbanisation and health. Julio's latest book is entitled *Urban Mobility and Poverty: Lessons from Medellin and Soacha, Colombia.*

Fabiola Berdiel is a PhD candidate in Public and Urban Policy and Director for International Field Programs at The New School. She is an urban researcher and community development practitioner. Her research interests include feminist and decolonial perspectives on urbanism and public space and design thinking as a tool for socio-economic development. Fabiola has a BA in Sociology and Political Science from Sarah Lawrence College and a Masters in International Affairs from The New School. Fabiola has worked and conducted field research in Guatemala, Nicaragua, Cuba, Argentina, Brazil, Mexico, Colombia, Puerto Rico, South Africa, Namibia, Ethiopia, Uganda, Senegal, Israel, Nepal, India, Kosovo, Serbia, Bosnia, Turkey, Ireland and Spain. She is co-founder of the research lab DEED: Development through Empowerment, Entrepreneurship and Design at The New School.

Maja Karoline Rynning is a doctoral researcher at the Institut Nationale des Sciences Appliquées de Toulouse (INSA), University of Toulouse, France. She is a civil engineer (M.Sc) and an architect (M.Arch). Her research explores the relationship between neighbourhood-scale built environment and people's mobility behaviours, with urban design as a strategy to promote a sustainable, zero-emission modal shift. Her work additionally addresses the importance of joint knowledge production research/practice combining evidence-based and experience-based knowledge to strengthen climate adaptation and mitigation through urban development.

Prologue

This book investigates how suitable research and planning could support a shift from pure *technical* focus on 'transport' to a more *holistic* vision of 'mobilities' in the Global South. Transport is distinct from mobilities as a concept. The term *mobilities* encompasses wider dimensions such as access to opportunities, life chances and wellbeing of people. The recognition of a changing society framed by the convergence of various forms of mobilities has been well established through understandings of 'sociology beyond societies' (Urry, 2000) and the 'new mobilities paradigm' (Sheller and Urry, 2006). The simple fact of traversing from A to B, which traditionally formed the basis of development planning (encompassing the domains of both urban and transport planning), is now transforming into multiple and complex trajectories between A to B via C and D, which heralds the emergence of multiple forms of mobility – or 'mobilities'. The ways in which mobilities get theorized and practiced have deterministic and perennial effects on cities' or regions' growth and livability (Sheller and Urry, 2006; Priya Uteng and Cresswell, 2007). Thereby, the produced, lived and enacted mobilities should not be understood as a mere concoction of physical movements alone but incorporate existing and new ways of physical/virtual; real/potential; isolated/combined; legal/illegal; formal/informal and empowering/disempowering movements of people (Priya Uteng, 2011).

The mobilities-turn has contributed to place this agenda to the forefront, and we need to deal with the challenges posed by the influx and intermingling of the varying formats of mobile systems and their implications for contemporary lives, as both researchers and practitioners.

Further, there exists some uncanny similarities between the developing countries when it comes to how a consistent focus is maintained on building infrastructures for car usage, albeit in the face of the continued dominance of non-motorized modes; blurred delineation between formal and informal modes both at governance and operationalizing levels; lack of analyses (resulting from a categorical lack of standardized data collection procedures, tools and methodologies) for context-enabled planning; and the social and spatial ramifications of difficulties related to movement of the majority. Additionally, the fact that the transport sector is among the top three contributors to greenhouse gas (GHG) emissions, the biggest consumer of non-renewable energy and has most negatively contributed towards urban pollution and traffic accidents makes it a suitable candidate for further analysis in the Global South context.

In the past 15 years, we have seen a consistent focus being developed on the thematic area of mobilities and its social ramifications in developed countries (e.g. Lucas, 2004; Lucas and Currie, 2012; Mejia-Dorantes and Lucas, 2014). There is a growing recognition and acceptance that movement/mobilities/accessibility/transport comprises an intersection of policies, infrastructure, dominant cultures and ICT technologies. But a similar focus is largely missing within developing countries. The ways in which this quadrilateral mix of policies/infrastructure/culture/ICT creates a variegated landscape of movements and interactions and affects different population groups remains understudied.

This book is an attempt to examine the ways in which different facets of the 'new mobilities paradigm' have converged to shape the lives of people in cities and regions in the Global South and post-colonial societies. It does so by examining the daily micromobilities (and immobilities) of people; histories of mobility; resource consumption (space, activities, space-time allocations); and the practices and power exercised by the state in governing mobilities. The book is situated in the cultures, spaces, forms and politics of mobilities, with a specific focus on planning practices. Corresponding to the new mobilities agenda, the book promotes and establishes new knowledge for transition towards integrated development systems in the Global South, through the following ways:

- Enhancing our understanding of inter-sectoral practices and addressing the potential effectiveness of measures to ease access to opportunities, with special reference to the marginalized sections of the developing countries.
- Exploring ways to provide a solid base for evaluating a wide range of measures, and relating them to the current policies in development planning, informed by both informal and (formal) state-driven practices.
- Thinking through consolidated strategies for increasing socially upward mobilities.
- Combining insights from several academic disciplines and research methodologies to address the field of mobilities.
- Putting forth ideas on how to involve stakeholders across various sectors to provide a platform for exchanging research and practical knowledge and enhancing applicability of the research findings.
- Promoting the uptake of new measuring tools and innovative methodologies for analysis (e.g. mobile phones and apps, which have a very high penetration even among the low-income groups in the developing countries).

The introductory chapter presents the cross-cutting theoretical, methodological and policy themes which are further explored in the following chapters.

Chapter 1 discusses the case of state-approved time/space restrictions on cycle rickshaws in the Indian capital, New Delhi, intended to improve conditions for car drivers. Economic and efficiency arguments by powerful authorities were effectively contested by grassroot coalitions mobilising support from academic, professional and other players.

Chapter 2 draws upon the case of Hanoi, the capital of the Socialist Republic of Vietnam, to present a different facet of 'mobility as resistance'. It highlights how state-led efforts to 'develop and polish' the city introduced a ban on street vending, while privileging auto-mobility. By remaining on the move, vendors maintained their livelihoods and claimed their stake to Hanoi's pavements. In this case, mobility itself became a tool for resistance.

Chapter 3 presents the interlinkages between daily mobility and mobile phones in two South African urban study sites. Increasing adoption of mobile phones and its penetration to the lowest income strata of developing cities provides new opportunities to unravel the travel behaviour/planning nexus. This has allowed new opportunities to both improve personal safety and to make better use of limited household funds for distance management by substituting virtual for physical travel whenever feasible.

Chapter 4 highlights an innovative way of data collection and usage, by capturing the ways in which technology (GPS tracks) and formal surveys can be combined to provide descriptive information and identify patterns of movement of the boda-boda drivers in Kampala. Informal transport solutions have a dominant and pervasive presence in the developing countries, but it remains difficult to access information about the lives of those involved in providing informal transport using formal approaches.

Chapter 5 presents another dimension of informality, by focusing on the affordances granted by Information and Communication Technologies (ICT) based solutions. It seeks to provide innovative ways to integrate the informal with formal transport services in ways which are rooted in the context and lived realities of the developing societies. The authors document and highlight the potential (positive) disruptive effects on transportation planning through mapping mass minibus transit systems in two case studies: The Digital Matatus project in Nairobi and the mapping of minibuses in Maputo.

Chapter 6 explores time use allocation and dialectics between mobile/immobile behaviour, using a three-week time use diary collected in Bandung Metropolitan Area (BMA), Indonesia. This survey is one of the first multi-day time use studies that has been conducted in the developing countries context. There were clear differences of weekday and weekend patterns of time use allocations and mobility behaviour across individuals and households from different socio-demographic groups.

Chapter 7 discusses the public transport system in Dar es Salaam its analysis of the regulation, organization and practices of public transport modes indicates how unrealistic it is to draw a clear distinction between formal and informal spheres. Such a dichotomy does not reflect adequately the various and often intangible forms of co-existence, merging and application of state regulation and of (implicit) rules and practices which stem from established and changing social norms, experiences or traditions.

Chapter 8 focuses on the spatial policies in South Africa, which historically and currently play a pivotal role in determining the existence and extent of mobility disadvantage in what is sometimes called the 'apartheid spatial form': low-density and decentralized cities with multiple 'nodes' of activity. The author highlights a common contradiction present across the developing world – the government *actively seeks* the informal sector to deliver services, and the existing regulatory and service gaps allow for the emergence of the private, and largely unregulated, informal sector, which eventually come to dominate the urban transport arena.

Chapter 9 has worked towards operationalising the concept of wellbeing of low-income urban citizens in the gambit of daily mobilities. Using the case of Abuja, Nigeria's capital city, transport-related vulnerabilities and disadvantages for low-income and vulnerable groups and how they are addressed by local transport is explored. The analyses have illustrated the importance of new knowledge(s) and methodologies critical to the future of transport planning in the Global South.

Chapter 10 offers a debate on a proposed conceptual framework urging us to focus on the agency, action and encounters of the urban actors and treat the "urban" as a place of 'in-betweenness' where the possible can be mobilized. This can be operationalised only when we are able to mobilise terminology, conceptual notions, images and diagrams that facilitate understandings of the diversity of localities, spaces and geographies. The pivotal role of "informal" solutions in rendering the city inclusive demands both increased recognition and inclusion in policymaking.

The Epilogue concludes the collection by offering an overview of the key issues that are raised by the country-specific case studies in order to draw some common narratives on the nature of the mobilities cultures in the Global South. It also discusses the importance of creating knowledge through dialogue between research and practice to reflect on the kind of overlaps or gaps that exist between research and practice. Generating this kind of knowledge is vital for designing and implementing context-specific sustainable mobilities solutions.

References

Lucas, K. (ed.) (2004) *Running on Empty: Transport Social Exclusion and Environmental Justice*. Bristol, UK: Policy Press.

Lucas, K. and Currie, G. (2012) Developing socially inclusive transportation policy transferring the United Kingdom policy approach to the State of Victoria? *Transportation* 39 (1), 151–173.

Mejia-Dorantes, L. and Lucas, K. (2014) Public transport investment and local regeneration: A comparison of London's Jubilee line extension and the Madrid Metrosur. *Transport Policy* 35, 241–252.

Priya Uteng, T. (2011) *Gender-Related Issues Affecting Daily Mobility in the Developing Countries*. Washington, DC: The World Bank.

Priya Uteng, T. and Cresswell, T. (eds.) (2007) *Gendered Mobilities*. Ashgate: Aldershot.

Sheller, M. and Urry, J. (2006) The new mobilities paradigm. *Environment and Planning A* 38, 207–226.

Urry, J. (2000) *Sociology beyond societies: Mobilities for the Twenty-First Century*. London: Routledge.

Some thoughts and acknowledgements

This book is partly a professional response and partly an emotional response to the phenomenon of *urbanisation* in the developing countries. I grew up in an Indian city which was yet to become *modern*, yet to face perennial traffic jams and yet to become the nightmare that it became. Cycle rickshaws and much later, auto-rickshaws catered to all our mobility needs. The needs were not much because everything was available in the local neighbourhood. A sustainable lifestyle from the perspective of urban and transport planning, so to say, was the only known modus operandi. There was a sense of wellbeing which comes from connecting to the local, being able to walk or cycle to the local and being the local. Then *Maruti Suzuki 800* happened in the late 1980s and everything changed. Almost overnight.

I am deeply grateful to Frode Longva and Tom Erik Julsrud at the Institute of Transport Economics, Oslo, for supporting this book. They could have asked me to simply put this thought aside and concentrate on more profitable projects. But they didn't.

I am also very grateful to my friends from DPS Bokaro 1996 batch. After approximately 21 years, they simply sprung up on my mobile screen one day, reminding me of the local lifestyle that once was. That reminder of the *local*, and a context which is no more, made me eager to give a more tangible form to the idea of this book.

A special note of thanks goes to Øystein Engebretsen and Aud Tennøy, my colleagues, and André Uteng, my colleague and husband, for being ever so creative and inspiring.

Karen, my co-editor, and all the contributing authors: Tusen takk!! (The literal translation is 'thousands of thank you', and in this case, I truly mean it.)

Many thanks also go to the Routledge publishing team, for being generous with their enthusiasm, patience and professionalism.

Tanu

The trajectories of urban mobilities in the Global South

An introduction

Tanu Priya Uteng and Karen Lucas

Introduction

Development projects often run the risk of being conceived and evaluated as stand-alone endeavors with streamlined 'positive' outcomes. Such optimistic rationales are often marred by unintended and unforeseen consequences. A long list of such consequences generated by urban and transport-related development projects exists.[1] The proportion is remarkably high in developing countries. Development and 'modernization' has left a long trail of, largely unaccounted for, costs. One of the undiscussed costs is related to the daily mobilities vis-à-vis life chances and wellbeing of people. Positioning this argument as its basic rationale, we examine the ways in which different facets of mobilities have converged to shape the urban areas in the developing countries and the ways in which existing mobilities are being modified.

Mobilities is a hidden concept and, to a great extent, an unknown term within transport policy circles both in the developed and developing parts of the world. The entire 'mobilities turn' discourse largely rests within the academic spheres of transport sociology and urban geography in the Global North. Nevertheless, we feel that it is also a potentially useful concept for unpacking the complex trajectories of urban mobilities in Global South and post-colonial cities. A key aim of the book, therefore, is to further bolster the mobilities concept within this context, especially in the domains of inter-sectoral research collaborations and 'research into practice' transport planning and policymaking.

Variations and inequalities in individual travel behaviours, along the lines of income, gender, race, age, working status, etc., are already a well-noted phenomenon in both the developed and developing world. However, the thematic areas falling under the umbrella of *travel behaviours* have often been studied without giving due attention to the causes and consequences of the highly unequal distributions of revealed, perceived and suppressed mobilities across different urban geographies and different population groups. This is especially true within Global South contexts where livelihoods and empowerment perspectives in the face of 'new' urban (neo-liberal) planning regimes and emerging mobilities have been insufficiently explored.

Very often the oversight of these important perspectives has resulted in urban social development programmes that are merely piecemeal 'sticking plaster' solutions, rather than a cohesive, coordinated attack on the root causes of problems of urban populations. This, precisely, is the case in the transport sector. Development-related projects may be robust in their technical details but equally ignorant of the nexus between enforced mobility and immobility, and with an alarming regularity, fail to make explicit the linkages between gender, livelihoods, mobilities and empowerment/disempowerment. Another common shortfall is the lack of context-specific and socio-economically sensitive planning, favouring a direct transfer of, often inappropriate, design and delivery standards from the West.

A further big divide within transport planning and delivery in many developing countries is the lack of recognition of the spatial variations of urban and peri-urban mobility and accessibility needs. The contexts at these three levels are significantly different from one another, and therefore, any generalizations would likely be wrong. The physical placement of people with low bargaining powers in accessing opportunities, space, activities and legitimization of this access (e.g. privatisation of public spaces; placement of Special Economic Zones (SEZs) in areas inaccessible by public transport, etc.) makes a deep impression on the ways mobilities are manifested in urban and peri-urban areas.

It is notable that the majority of the current slum resettlement programmes and low-income public housing schemes in developing cities actively relocate the poorest to the most peripheral parts of the city, where they have to transverse the worst barriers to inclusion to access its opportunities. Whether the transport planning context is intervention or neglect, it is always the poorest populations that face the greatest challenges as they attempt to bridge the gap between physical and economic allocations. Forced to live in the least accessible (lowest cost) peripheral locations, far from major public transport routes, with the least resources to purchase personal transport or to pay transport fares, they are further marginalised by lack of power and voice (Ahmed et al., 2001; Naumann and Fischer-Tahir, 2013). Long working days, sandwiched between long, often uncomfortable, occasionally dangerous, journeys to and from work, reinforce the friction of distance.

This is a relationship full of contradictions – on the one hand mobility helps to raise quality of life standards by offering improved access to city opportunities, whilst on the other hand, the ubiquitous lack of access to transport services amongst the urban/peri-urban poor severely constrains their potential for economic and social development (Tiwari, 2003; Ahmed et al., 2001; Mandri-Perrott, 2010; Cervero, 2013; Naumann and Fischer-Tahir, 2013; Oviedo and Dávila, 2016).

Another vital, and rarely discussed question emerging from observation of these current trends is how do the forces of globalisation, urbanisation, changing livelihood strategies and mobilities intersect? Because of the rise and intensity of risk and uncertainty associated with climate change, disaster and conflict situations, and the precarious position of low-income populations in these situations, it is important that mobility needs assessment are seen as important components of all development decisions and processes.

The urgent need for socially stratified understandings of mobilities in the context of Global South cities is further augmented by the fact that more than half of the world's population is now living in urban areas. The urban share is likely to rise from 75 per cent to 81 per cent within developed countries between 2007 and 2030, and from 44 per cent to 56 per cent in developing countries (UNFPA, 2007). Present forecasts predict that urbanisation will occur most rapidly in Africa and Asia, doubling its urban share between 2000 and 2030. Apart from these population trends, 'globalisation' in the developing world has largely meant braving the spillovers of various development activities of the Global North.

In the following sections, we briefly discuss a few overarching issues that are further explored in the book, and which bear significant repercussions for the mobilities narrative. We also seek to explore how current planning regimes and institutional framings are reinforcing and reproducing planning policies and designs which have been acknowledged as mistakes in the developed countries. The chapters in this book highlight the interplay between the organisation of both formal and informal planning institutions; policies and programmes; social practices; emergence of a powerful middle class and the associated status representations; changing contours of development politics; multiculturalism; and (as)symmetries of power and opportunities; pervasive existence of informal sector; conflict situations and displacement; and issues related to governance (Absar, 2000; Houpin, 2010; Mlay R. et al., 2010).

The mobilities concept is particularly useful as an analytical lens for exploring these important because it specifically brings the asymmetries of power and denied opportunities to the fore. A common theme permeating all the chapters within the book (drawing its essence from Cresswell's definition [2001b,2006]) is that movement itself, or a lack thereof, does not possess any inherent meaning, but needs to be understood and elucidated as a socially constructed and highly contextualised phenomenon reproduced through both material and discursive representations and practices.

Mobilities

Discourses on the concept of mobility have traditionally described it as physical movement (operating in the domains of geography, urban planning and transport) on one hand and a change in social status on the other (a sociological construct). The merging of these two realms in the research field is a recent phenomenon, but a segregated approach is still visible in the field of planning practice and policymaking. A slow but steady progress towards including dimensions of needs, liveability, walkability, wellbeing and social exclusion in the Global North can be traced from the mid-'90s onwards. Developing countries have largely remained aloof to such interlocking mechanisms, and the field of transport remains segregated from understandings built on social feedback loops.

Unfortunately, this trend has continued to rise unabated. Addressing this issue will involve looking into space, place, locality and mobilities as a physical, cultural and social category (see Gregory and Urry, 1985; Featherstone, 1990; Lash and Urry, 1994; Urry, 2000; Urry, 2007; Latour, 1999; Bonß and Kesselring, 2001, 2004).

The cohesive nature of social and spatial mobility implies that a change in geographical/spatial mobility patterns affects the individual space of options and action, thus producing varying terrains of social mobility. These understandings metamorphosed in the introduction of a 'new mobilities paradigm' (Sheller and Urry 2006a; Hannam Sheller and Urry 2006), which posits that *movement*, *representation* and *practice* are embedded in uneven socio-political relations (Creswell, 2006). While outlining this paradigm, Urry (2004, 28) emphasises the need to separate out, rather carefully, the nature of the five highly interdependent 'mobilities' that form and reform social life, bearing in mind the massive inequalities in structured access to each of these:

1. Corporeal travel of people for work, leisure, family life, pleasure, migration and escape.
2. Physical movement of objects delivered to producers, consumers and retailers.
3. Imaginative travel elsewhere through images of places and people on television.
4. Virtual travel often in real time on the Internet, so transcending geographical and social distance.
5. Communicative travel through person-to-person messages via letters, telephone, fax and mobile phone.

What this composite set of activities and interactions brings to the fore is that people's mobilities (as well as immobilities – in terms of both the right of people to remain free from the demands of excessive or enforced mobility, as well as the denial of resources to allow them to secure adequate mobility for their livelihood needs) must be studied, interpreted and theorised in embodied and contextualised experiences of movement (e.g. Löfgren, 2008), through discursive representations (Cresswell and Priya Uteng, 2007; Mountz, 2011) and through discerning its relationships with inequality and governmentality (Ohnmacht et al., 2009).

On a similar note, Langan (2001: 459) describes the theme of mobility as a desired end that is not only a function of personal achievement but also a product of several constituent and affecting parameters:

> Rousseau long ago declared in *The Social Contract* that the cripple who wants to run and the able-bodied man who doesn't will both remain where they are. But by focusing on internal resources and intentions, Rousseau forgot to mention all those whose mobility is affected by external constraints. To consider those constraints is to notice how the built environment – social practices and material infrastructures – can create mobility-disabilities that diminish the difference between the "cripple" and the ambulatory person who may well wish to move.

Concurring with the importance of context in the production of mobility, Cresswell (2001b) espouses mobility as a movement that is socially produced, is variable across space and time, and has visible effects on people, places, and things, and the relationships between them. Jones (1987: 34) puts forth the three components of *individual action, potential* and *freedom of action* to express mobility. In short, he interprets these as:

- *Individual action*: in the form of observed movement or travel;
- *Potential action*: in terms of journeys that people would like to make, but are unable to because of limitations in the system and/or their own commitments restricting them in time and space, or financial restraints; and
- *Freedom of action*: which may never manifest in action, but gives the individual options from which to select and the knowledge that he/she could do something.

Knie (1997) introduces a related understanding of the mobilities concept, emphasising that mobility is about the *construction of possibilities for movement*, rather than actual movement itself. Sørensen (1999) notes that the analysis of mobility is basically about *the performance, real as well as symbolic*, of the provision of physical movement in society.

Importantly, Kaufman (2002) puts forward the idea of mobility as a *restricted good*. These opportunities emerge as a function of market relations; for example, being related to and being limited by geographical locations. In the following points, he (2002: 37) identifies the three determining factors shaping the mobility levels and patterns of individuals:

- *Access* to mobility-scapes (representing transport and communication infrastructure as potential opportunities).
- *Competence* referring to the 'skills and abilities' necessary to use the accessible mobility-scapes.
- *Appropriation*, as a third factor, involving all behavioural components, such as the need and willingness to make use of the scapes, to become mobile.

As such, he implicitly suggests that individuals with different levels of competence and access to power will have different (and inequitable) access to mobility resources, with significant consequences for their access to financial and social capital. To make these social consequences manifest, Nijkamp et al. (1990: 22–24) recommend that the analysis of mobility and the underlying causes for its demand should be undertaken at a broad scale in the context of the following four themes:

- *Socio-economic context* of analysis, which focuses attention on the influences of exogenous socio-economic conditions upon spatial patterns of interaction;
- *Technological context* of analysis, which deals with the implications of changes in the technological environment on the spatial behaviour of individuals or groups in our society;

- *Behavioural analysis*, which focuses attention on motives, constraints and uncertainties facing individuals, households and groups when taking decisions regarding transport, communication and mobility; and
- *Policy analysis*, which concerns the evaluation of actions, usually the application of policy instruments or measures of decision-making agencies regarding transport.

Sorensen (1999) summarises these thoughts, under the term 'mobility regimes', in order to highlight the historical and cultural basis of mobility. A mobility regime results from a number of factors, some of which consist of the physical shaping of cities and landscapes, the available transport and communication systems, the relationship between mobility and economic, social and cultural activities and the meaning attributed to mobility.

These reflections on the theoretical insights taken directly from the mobilities discourses suggest that mobility cannot be analysed in a purely instrumental, objectivist mode. Rather it retains a subjective dimension differing with the distribution pattern of its constituent resources (e.g. time, money, infrastructure, opportunities, etc.). Differential accessibility to these mobility resources maps out different mobility regimes distinguishable at the levels of people, places and processes. *Mobility* thus emerges as an enabling characteristic, a sought-after rather than given 'good/commodity'. The understanding of mobility has thus entered the wider realm of discussion on identity formation, freedom and rights to the city (Fainstein, 2014).

Mobilities also need to be understood through an equally concentrated approach towards understanding 'immobilities'. The mobilities approach facilitates this by stressing the need to understand the relationship between fixity and movement. It pushes us to understand if, why and how, choices are made to remain stationary, that is, do these outcomes arise out of preferred choice sets, or are the immobile people prey to circumstantial gameplays? As such, mobility cannot be ascribed with positive connotations alone, and as Bergmann and Sager (2008: 2–3) highlight:

> with modernity firmly fixed to rapid movement in numerous forms, it is [the] responsibility of social researchers to explore potential alternatives to travel. Is there a 'thrill of the still' – a viable slow city strategy, for example – to counter the ideals inherent in terms like the jet set, living on the fast track, fast cars/fast food, etc.?

Unpacking the relationships between society, mobilities and politics, or rather understanding the 'socio-political dimension of mobility', highlights the nexus between mobility and immobility. It further assists in figuring how such connections are being made, if these connections are being taken into account while making development decisions, and in absence of any interventions, which mechanisms are the affected population adopting? This will allow decision-makers to chart the livelihoods and social setups that might result from new infrastructures and other related land use and service interventions. For example, for policymakers to see the gendered effects of large transport developments, one can borrow from the research carried out by mobilities scholars to understand how moving between borders affects producing, effacing or re-affirming social differentiations (Hyndman, 1997; Hyndman and Mountz, 2006; Amoore, 2006; Mountz, 2011).

Similarly, the idea of appropriation (Mountz, 2011) and affordances (Gibson, 1982), crucial elements for mobilities, are directly linked to the issue of governing a heterogenous city. This requires revisiting the state policies and associated outcomes. Migrant poor (often labourers), for example, are not recognised as citizens of civil society per se, but part of the political society and often used as vote banks in the wake of election campaigns. What this effectively means is that the migrant labourers are never considered as stakeholders in the mobility planning design processes per se. Amin (2013) invokes similar sentiments while introducing 'telescopic urbanism'. He warns against the large-scale engineering of wellbeing where the majority of the city, particularly in developing countries, disappears while drawing urban development plans and programmes. A singular focus on the elite, which in this case can be interpreted as elite mobilities, needs to be dissolved and we need to turn around the telescope so that the entire city comes into view.

Several chapters in this book highlight how new mobilities are arising from state interventions within the transport domain. For example, putting time/space restrictions on cycle rickshaws in the Indian capital, New Delhi (Chapter 1), plays a crucial role in enforcing multiple levels of exclusion. Enforcing workers to search for alternate livelihoods; changing social roles and norms within the family or society at large; wholesale relocation of communities, etc., necessitate understanding how geographies of inclusion and exclusion are being carved out because of changing contours of mobilities.

The thematic areas of mobilities occupy a non-negotiable position when disusing policies directed towards social development. The following sub-sections sketch how this position is further explored in the book.

The spatial content of mobility

Understanding the spatially differentiated mobilities of different social groups living within developing urban contexts can lead to better formulations of its space-making and land-use attributes. Space, whether sacred or profane, is not produced in vacuum, but rather through a web of crosscutting power relations that are themselves forged at multiple scales from the local to the global (Massey, 1993; as quoted in Secor, 2002: 7).

To build up a mobility profile of the developing world, the research arena provides a good foundation to enumerate the differences and build a case for redesigning of methods, analyses and policies. Further, the issue of transport-related social exclusion and constrained mobilities is also highly dependent on the way cities are being produced. The cases presented in this book illustrate how downgrading of informal transport systems, the interplay of time-space distributions, the relationship with employment opportunities, less flexibility in combining various spatial activities, etc., all create and perpetuate vicious cycles for the urban poor and most socially vulnerable in society.

In this book, we argue that unless planned interventions include the 'lived realities' of the urban poor, transport interventions will continue to create and exacerbate social exclusion. It is, therefore, imperative to discuss the distribution of resources by filtering through the following evaluation stages:

- Mobility resources: what is available?
- Access to important nodes of activities: what is the extent of current coverage? How can existent and newly emerging mobilities be strengthened for accessing new opportunities?
- Use of the current modes, including the informal ones: how effectively are resources used?
- Capacity: what is the capacity to manage resources?
- Environment: what are the environmental impacts?
- Deciphering neoliberal practices: how does it impact daily mobilities?
- Are mobility plans drawn while making decisions on community development, education, health and employment?

Although there is still a paucity of research to unravel these multiple facets of mobilities within developing contexts, a number of studies have highlighted the dissociations between mainstream transport planning and the existing needs/conditions in the developing countries (Lucas, 2009; 2011; Priya Uteng, 2011; Maia et al., 2016; Porter, 2016; Lucas and Porter, 2016).

An illustrative case is the emergence of the 'new middle classes' in many cities of the Global South. A steep growth in the volume of an economically well-off and influential middle class in the past two decades has sketched new contours of a 'developed society' within the development contexts. These symbolisms are often noticed in the emergence of a new urban geography, indicated by shopping malls, leisure complexes, high-rise office blocks, gated housing areas and successive progressions towards 'better' car brands. Through their sheer dominance in framing the urban landscapes, this class has also eventually come to dictate how daily mobilities need to be understood, (re)produced and maintained across different scales through forms of labour (mostly fixed origin-destinations), consumption (ruled by the emergence of shopping malls), sociability (facilitated by car-based mobility) and politics (development defined in strictly Western terms).

Often these symbolic powers are taken as the cornerstone for planning daily mobilities in the city, and increasingly for locking out non-motorized modes of both private and public spaces. So, for example, three-wheelers or cycle-rickshaws are not allowed to operate on certain flyovers or allowed to enter the high-end hotel premises. Some of the institutional consequences of the emergence of a rich and powerful middle class have resulted in the emergence of a strong political mandate for highway- and road-building agendas, supported by a powerful car manufacturers lobby. These so-called development projects are then valorised at the time of elections, as the physical tokens of progress made by the ruling government.

What emerges in terms of analysing the decision-making processes and the resultant spatial policies and programmes is a conceptual and policy blackhole. Seldom are the following questions asked and studied:

- How is the middle class shaping the spatial mobilities in the urban and peri-urban areas? Which groups are being excluded?
- How are the new forms of spatial inequalities arising in the middle-class cities?
- What is the role being assigned to the existing forms of 'shared mobilities' in the developing cities?

The social content of mobility

The interaction between spatial mobility for negotiating daily lives and other forms of mobility (social, economic and political) has not been substantially explored in the developing countries. Specifically, a focus on the operationalisation of social norms (and consequently, mobilities) both permits and requires inquiries into the themes of development, democracy, equity and their distributions in different societies. Enlarging this understanding, it also becomes essential to evaluate the differential claims of cultural beliefs and norms as being equally, or in some cases the most, important factor dictating mobility, along with how these beliefs and norms are changing in the present times. Are these changes supportive or detrimental to the cause of inclusive development and empowerment? Which issues can be corrected through policy interventions to facilitate positive changes?

Further, it is widely assumed that 'the convergent effects of globalization and cross-border organizational learning have rapidly outpaced the divergent effects of cultures, national institutions and social systems' (Mueller, 1994). Products of unrestricted mobility and the markers of postmodern times, time-space compression and social fluidification have come to be accepted as given characteristics of the present times. Yet both these concepts remain grossly under-examined in terms of their social distributional effects.

Kaufmann, in an analysis of social fluidification, notes that 'the crux of the debate is whether or not the compression of time-space goes hand in hand with a decrease in certain social constraints that discourage action. It is, thus, a question of analysing who has access to which relevant technologies and the degrees of freedom afforded by their usage (Kaufman, 2002:14). The very idea of fluidification supposes that social and territorial structures take the backseat to a context that is capable of accommodating the most diverse aspirations (ibid.: 87). But is this true in fact? Massey, in a critique of Harvey's all-encompassing notion of time-space compression, remarked that 'different social groups have distinct relationships to this, anyway differentiated, mobility; some are more in charge of it than others; some initiate flows and movement, others don't; some are more on the receiving end of it than others; some are effectively imprisoned by it' (Massey, 1993: 61). Cresswell succinctly captures this point of view:

> The question of how mobilities get produced – both materially and in terms of 'ideas' of mobility – means asking: who moves? How do they move? How do particular forms of mobility become meaningful? What other movements are enabled or constrained in the process? Who benefits from this movement? Questions such as these should get us beyond either an ignorance of mobility on the one hand or sweeping generalizations on the other.
>
> (Cresswell, 2001b: 25)

A similar line of research under the aegis of the Social Exclusion Unit (2003) in the UK was promoted under the theme of 'transport and social exclusion'. Certain key points can be gleaned from this research agenda to enlarge the discussion on differentiated mobilities, such as categorisation of excluded groups, time-space interplay and its differential structure, and place/social-category/person-based measures. This position can be further strengthened by analysing the dialectic and operationalisation of moorings and mobilities (Urry, 2003: 128), through answering:

- How are mobilities encoded in the belief systems? Which movements are prescribed, and how can these be harnessed for empowering different groups?
- Under which circumstances and through which mechanisms are mobilities being produced and maintained?
- What does governance mean in cases where new mobilities emerge as a direct outcome of the state's development decisions (including the coalitions between multi-national companies and government authorities)?

On a micro level, Jacobs' (1961) seminal view of street as a site of social interaction as much as a space of circulation needs to be bolstered. 'People's positions are mediated by their habitual activities in moving about the city. The common practice of walking, bicycling, bus-riding, or driving constitute distinctive forms of urban life, each with characteristic rhythms, concerns, and social interactions' (Patton, 2004: 21). Thus, as Jensen (2009) notes 'what should be acknowledged is therefore the dialectic relationship between place and flow, between the global movements and the local relationships (e.g. Massey, 2005; Morley, 2000)'.

The right to the city (RTC), the New Urban Agenda (NUA) and the ground realities

Another issue that is highlighted throughout this book, and which reflects current and emerging planning research and policy practices in the world at large, is the pressing need to engage with an increasingly urban world. The concept of 'Right to the City' (RTC) invoked in the New Urban Agenda (NUA) is the latest attempt to this end. The basic postulates surrounding the agenda urge us to revisit the definition of *urban*. Further, which groups are enabled to exercise their RTC given that these rights are, at best, a diffuse set of categories, and the costs involved with exercising these rights.

Elaborating on the topic of 'urban', it is evident that the urban areas are primarily emerging as 'fractals of peri-urban interfaces' rather than pure urban forms. This implies that the geography, physicality and the loop of resource consumption and production that supports urban life is constantly shifting. This has clear repercussions for both the claims being made on the urban areas vis-à-vis RTC, and the associated responsibilities. The most recent response to this dilemma is evident in the emergence of the NUA. Habitat III (the United Nations Conference on Housing and Sustainable Urban Development, 2016) makes an explicit commitment to one of the Sustainable Development Goals (SDGs) – SDG number 11 (Sustainable Cities and Communities) – and enshrines RTC as the foundation of urban development in creating 'just and inclusive' societies. The NUA, however, remains largely silent with respect to implementation.

As the status of both the RTC and 'inclusive urban development' remains conceptually vague and open to interpretations, this is hardly surprising. Often these two concepts are mobilised by researchers as a slogan, as an analytical category, and sometimes both at the same time (Houssay-Holzschuch, 2017 expands on this while discussing RTC). Amin (2013: 487) poses the question 'what of the substantive minima of a rights-based approach to urban inclusion, the "obligatory points of passage" through which the poor find the hand up to a fulfilling life?'

Implementing the goals of the NUA requires direct confrontation of issues related to differentiated access to opportunities, power and decision-making. This rests on a logical framing akin to what Houssay-Holzschuch (2017) bundles under the 'De facto RTC'. The De facto RTC concentrates on the fault lines between processes of spatial and social ordering and public action (design and practice of public policies), routines of urban daily practices and construction of a socio-spatial order via occupying and appropriating space, and consolidation of social connections. It also questions the ways in which processes informing the urban, spatial, political and social order get institutionalised in the long term.

Unbundling such a complex nexus requires careful mapping of the different development sectors. The NUA responded to the institutional dimension (of the local) by outlining the following demands:

- Local and regional governments should have a seat at UN negotiation tables and be able to take decisions without the interference of national governments.
- Local and regional governments should have direct access to international finance. This implies that 20 to 25 per cent of global finance for development – in instruments such as the Green Climate Fund – are to be considered for direct allocation to the cities.

This can be brought to fruition only if an 'arrangement of needs assessment' is introduced, which can effectively monitor change and act accordingly. Thus, the core idea that the local urban governments and those within their jurisdictions whose needs are not met (including representative organizations of the urban poor) be actively involved in the implementation of the NUA, needs a systematic framework of operationalisation.[2]

> By invoking the issue of deep (re)distribution, we can redress the structural causes of poverty, injustice and exclusion, and adopt affirmative actions to counteract them. Though these ambitions are not new, the unpacking of Habitat III Urban Agenda presents a unique opportunity to address exclusion, unsustainability, and the ways in which they are being reproduced at multiple levels of resource consumption. Further, it provides an opportunity to examine the quality of urban governance needed to address the existing and evolving patterns of exclusion and unsustainability..
>
> (Priya Uteng, 2016: 21)

However, simultaneous to these discussions of poverty reduction and a fairer distribution of resources within cities, there is a strongly emerging technocratic discourse of The Smart City, which is fast gaining ground in the Global South. India, for example, has one of the most ambitious smart cities programmes. The smart cities agenda tends to shift the policy focus away from the mobility needs of people (and especially more vulnerable and disenfranchised populations) onto the need for more intelligent buildings, agility in distributed energy systems, intelligent streets, smart parking, public transit tracking and other technical responses. Any reflection on issues of exclusion, social justice and 'cities for all' simply do not feature in this essentially corporate-driven smart cities agenda. Questions such as whether urban dwellers will be reduced to consumers/users or still count as citizens; which characteristics will define a smart citizen; whether smart cities can be good for everyone and how will issues such as affordability, social justice and access to opportunities be coded in the smart algorithms are yet to feature in the smart cities agenda.

This book suggests that these technocratic approaches to mobilities will not work in the context of Global South cities, and/or will only make things better for a minority elite within these cities. We need to be careful about such elite imaginaries of the future becoming the dominant norm. The apparent apathy towards the daily mobilities within the smart cities agenda is primarily due to a general lack of understanding on how to manage the complex issue of affordances granted to different groups to access vital activities and its ties to transport/land-usage development. The absence of any successful model of how to cope with rapid urbanisation and woefully inadequate existing infrastructures found in virtually all Global South cities is merely a manifestation of this endemic situation. These conditions have evolved into 'wicked' systemic failures, and though cities are developing, they are developing chaotically, without the requisite supporting infrastructures, resources and overarching governing regimes. Risk of exposure to traffic fatalities/injuries and lethally high traffic-related pollutants in the developing world, for example, is not an isolated event given the voluminous incidences, but rather has become a pandemic systemic condition of everyday mobilities, which are deeply embedded in, and undermining almost every aspect of the daily lives of citizens in Global South cities and their sprawling hinterlands.

How capable is the discipline of urban transport planning in the developing countries?

Serious questions can be raised about the capabilities of the urban transport planning sector in developing countries. Although urban transport is as much a cultural, social and political issue as it is technical, the collusion of politics with technical rationality in justifying decisions to support and promote a car-based culture takes the front seat in practically all development-related discussions. The very ethos of development in the Global South is currently being rolled out through a series of road- and bridge-building programmes.

The primacy of these technically driven solutions often obscures grounded discussions on issues on who moves and who does not, and why certain groups of people are mobile while others are not. Is it simply a matter of choice or of failed policies? Determining who precisely benefits and disbenefits from the spatial decisions taken at the regional and local levels is also largely not considered. In fact, how mobility and immobility affects different groups of people along the lines of gender, age, social class, employment tenure and other aspects of socio-spatial disaggregation is all too often entirely overlooked. As such, important questions of how livelihoods are affected by being mobile or immobile and how people get in and out of transport poverty almost never get asked.

Even a cursory analysis of the transport statistics in the developing world will reveal that, although the modal split primarily tilts towards sustainable modes such as walking, cycling and public transport, the sheer volume and growth in motor vehicles, coupled with traffic accidents in the urban realm, brings any policy discussions of sustainable mobility to an abrupt halt. The existence of corruption is often portrayed as the main handicap affecting policymaking in developing cities. However, the capacity of local decision-makers and urban/transportation planners in these contexts is also seriously restricted by their lack of knowledge about the variety of methods that are available to answer the relevant questions of how to provide socially sustainable transport systems. One important issue from the perspective of this book is how can the research community ensure that these questions get asked?

Looking forward

Studies across a host of countries in the developing world are consistent in their findings that the current transport policies that are being adopted do not place an appropriate level of importance on the lived realities of daily mobilities (Fouracre et al., 2006; Lucas, 2011). Rather, they are mainly concerned with the design and operations of transport systems as purely infrastructure systems, and much less involved with issues of daily mobilities in terms of access to livelihoods, education, health, etc. Attention to these alternative lines of enquiry have been much more commonplace within anthropology, psychology, geography and sociology (e.g. Bebbington and Batterbury, 2001; Scoones, 2009; De Haan, 2012, Bocarejo and Oviedo, 2012; Delmelle and Casas, 2012; Oviedo and Dávila, 2016) than the engineering and planning disciplines.

This book attempts to broaden the discussion on urban mobilities in the developing countries. It is based on a firm belief that a sole focus on technocratic approaches in the transport field, without taking into cognizance the structural differences in access to resources, social norms and issues related to safety, has diluted the effectiveness of policies and plans. It builds on cases to align a theoretical framework for understanding and commenting on mobilities and to further indicate possible methodological and policy directions. Cases are presented to readdress the understanding of 'development'. Development projects, it is argued, should intrinsically have an inbuilt agenda on mobilities – detailing out 'mobility needs' and 'mobility gaps' – at its inception stage. These cases also serve to highlight ways in which monitoring can be made an essential part of development projects.

Despite efforts by various national and international development organisations, equal access to mobility opportunities remains a problem in the developing countries. A further countervailing issue is that policy- and decision-making in the transport sector is based on dynamics of power that are often in generous interaction with corrupt routines. This is happening despite a wide policy acceptance that the most sustainable forms of transportation are public transport, cycling and walking. Paradoxically, it can be noted that the majority of women, children, the elderly and the disabled, as well as almost all the poorer segments of the population in developing cities, are already practicing these sustainable forms of mobility. Unfortunately, this is not borne out of choice but because of their inability to secure private motorised mobility resources. Instead of creating an environment where these groups can continue with their sustainable mobility practices, they are increasingly forced into unaffordable, unsafe and unregulated (or partially regulated, informal) motorised modes.

Through this book, we would like to advocate that future planning interventions in Global South cities can, and should, be built to sustain these current mobility practices, but under more conducive economic and social conditions. This can be achieved through designing policies aimed at increasing the patronage of public transport, cycling and walking by improving the systemic conditions and infrastructures for the use of these modes. We recommend that people- and needs-based approaches be taken more seriously within the transport planning domain if socially sustainable mobility futures are to be achieved by developing cities.

Questions regarding the tools, methods and results that help the decision-makers make informed decisions are critically important. Research studies grounded in the 'new mobilities paradigm' have drawn upon a wide variety of methods, including non-standard data collection and analytical approaches, but with a strong emphasis on mixed-methods and interdisciplinary approaches. A mix of data collection techniques such as cognitive mapping exercises; semi-structured interviews combined with GIS/GPS-analysis of local travel patterns; structured interviews with spatial visualisations; photographs and observational research; data from technical appraisals of the transport system; app-based and mobile phone tracking–based mapping and data gathered on subjective wellbeing, etc., can facilitate this shift. As the chapters herein demonstrate, these methodologies can provide a rich source of data and a more reliable platform to build people-centred understandings of mobilities. Reimagining and rethinking 'transport' by posing the right questions and adopting a mix of approaches to discuss the dynamics of the symbolic, political and actual role played by transport infrastructure in the daily lives of people, provides a good vantage point for switching to 'mobilities'.

Notes

1 'For example, the case of the LA Bus Riders Association (Soja, 2000), the social construction of the 'tramp' (Cresswell, 2001a), graffiti as subversive and political action in the street (Cresswell, 1996), the socially segregating effects of the Bangkok Sky Train (Jensen, 2007), the phenomenon of transportation racism in America (Bullard et al., 2004) and the contested forms of automobility (Bo¨hm et al., 2006). Some authors notice the bottom-up character of some types of mobile politics and speak of 'grassrooting' the practices of mobile politics (Castells, 2005), and how an intricate relationship among bodies, networks, flows, identities and protest surfaces in the People's Global Act (PGA) in India (Routledge, 2005).' (Jensen 2009: 148).

2 While discussing human rights law and the city, Clément (2017: 12) notes that 'Many social problems require far more systemic solutions than the legal system can provide. In no country have the courts been capable of redistributing resources to alleviate disparities in wealth, even in India and South Africa where there are constitutional guarantees for social and economic rights. . . . But cities are far better off promoting alternative ways of addressing issues such as poverty and racism than litigation before human rights tribunals.' If the NUA promotes a financial (and fundamental) independence of the cities, it is vital that cities start acting fast to unravel the connections between the different sectors and attacking the structural loopholes in development, to avoid legal challenges once the RTC gets enshrined in the laws dictating cities. These (future) legal implications are obvious but poorly understood. As Agrawal (2017) states the same for Canada while discussing human rights and the city, 'Despite such a historically noteworthy legislative environment for human rights in this country, many Canadian scholars and practitioners in planning are still unsure – or worse, unaware – about how these constitutional and quasi-constitutional requirements apply to planning matters at the municipal level.'

References

Absar, S.S. (2000) Conditions, concerns and needs of garment workers in Bangladesh. *Development Bulletin* 51, 82–84.

Agrawal, S. (2017) Human rights and the city. *Plan Canada* 57 (2), 4.

Ahmed, S.M., Chowdhury, M. and Bhuiya, A. (2001) Micro-credit and emotional well-being: Experience of poor rural women from Matlab, Bangladesh. *World Development* 29 (11), 1957–1966.

Amin, A. (2013) Telescopic urbanism and the poor. *City* 17 (4), 476–492.

Amoore, L. (2006) Biometric borders: Governing mobilities in the war on terror. *Political Geography* 25, 336–351.

Bebbington, A.J. and Batterbury, S.P.J. (2001) Transnational livelihoods and landscapes: Political ecologies of globalization'. *Cultural Geographies* 8, 369–380.

Bergmann, S. and Sager, T. (2008) Introduction. In: Bergmann, S. and Sager, T. (eds.), *The Ethics of Mobilities*. Aldershot: Ashgate.

Bocarejo, S.J.P. and Oviedo, H.D. (2012) Transport accessibility and social inequities: a tool for identification of mobility needs and evaluation of transport investments. *Journal of Transport Geography* 24, 142–154.

Bohm, S., Jones, C., Land, C. and Paterson, M. (eds.) (2006) *Against Automobility*. Oxford: Blackwell.

Bonß, W. and Kesselring, S. (2001) Mobilität am Übergang vom Ersten zur Zweiten Moderne. In Beck, U. and Bonß, W. (eds.), *Die Modernisierung Der Moderne*, 177–190. Frankfurt: Suhrkamp.

Bonß, W. and Kesselring, S. (2004) Mobility and the cosmopolitan perspective. In: Bonß, W., Kesselring S. and Vogl, G. (eds.) *Mobility and the Cosmopolitan Perspective*, Workshop at the Munich Reflexive Modernization Research Centre (SFB 536), 29–30 January.

Bullard, R.D., Johnson, G.S. and Torres, A.O. (eds.) (2004) *Highway Robbery. Transportation Racism and New Routes to Equity*. Cambridge, MA: South End Press.

Castells, M. (2005) Space of flow, space of place: materials for a theory of urbanism. In: Sanyal, B. (ed.), *Comparative Planning Studies*, 54–63. London: Routledge.

Cervero, R.B. (2013) Linking urban transport and land use in developing countries. *Journal of Transport and Land Use* 6 (1), 7–24.

Clément, D. (2017) Legally speaking: Human rights law and the city. *Plan Canada* 57 (2), 10–13.

Cresswell, T. (1996) *In Place/Out of Place: Geography, Ideology and Transgression*. Minneapolis, MN: University of Minnesota Press.

Cresswell, T. (2001a) Making up the tramp: Towards critical geosophy. In: Adams, P.C., Hoelscher, S. and Till, K.E. (eds.), *Textures of Place: Exploring Humanist Geographies*, 167–185. Minneapolis, MN: University of Minnesota Press.

Cresswell, T. (2001b) The production of mobilities. *New Formations* 43, 11–25.

Cresswell, T. (2006) *On the Move: Mobility in the Modern Western World*. London: Routledge.

Cresswell, T. and Priya Uteng, T. (2007) Introduction to gendered mobilities. In: Uteng, T.P. and Cresswell, T. (eds.), *Gendered Mobilities*. Ashgate: Aldershot.

De Haan, L.J. (2012) The livelihood approach: A critical exploration. *Erdkunde* 66 (4), 345–357.

Delmelle, E.D. and Casas, I. (2012) Evaluating the spatial equity of bus rapid transit-based accessibility patterns in a developing country: The case of Cali, Colombia. *Transport Policy* 20, 36–46.

Fainstein, S.S. (2014) The just city. *International Journal of Urban Sciences* 18 (1), 1–18.

Featherstone, M. (1990) *Consumer Culture and Postmodernism*. London: Sage.

Fouracre, P.R., Sohail, M. and Cavill, S. (2006) A participatory approach to urban transport planning in developing countries. *Transportation Planning and Technology* 29 (4), 313–330.

Gibson, E.J. (1982) The concept of affordances in development: The renascence of functionalism. In: Collins, W.A. (ed.), *The Concept of Development: The Minnesota Symposia on Child Psychology* (Vol. 15). Mahwah, NJ: Lawrence Erlbaum Associates, Inc.

Gregory, D. and Urry, J. (1985) *Social Relations and Spatial Structures*. London: Palgrave Macmillan.

Hannam, K., Sheller, M. and Urry, J. (2006) Mobilities, immobilities and moorings. *Mobilities* 1, 1–22.

Houpin, S. (2010) *Urban Mobility and Sustainable Development in the Mediterranean - Regional Diagnostic Outlook*, United Nations Environment Programme Mediterranean Action Plan - Plan Bleu Regional Activity Centre.

Houssay-Holzschuch, M. (2017) CFP: The right to the city in the South, everyday urban experience and rationalities of government, Paris, November 2017 (call for papers). Available at: https://networks.h-net.org/node/22277/discussions/170392/cfp-right-city-south-everyday-urban-experience-and-rationalities (accessed 20 March 2017).

Hyndman, J. (1997) Border crossings. *Antipode* 29, 149–176.

Hyndman, J. and Mountz, A. (2006) Refuge or refusal. In: Gregory, D. and Pred, A. (eds.), *Violent Geographies*, 77–92. London, Routledge.

Jacobs, J. (1961) *The Death and Life of Great American Cities*. New York: Random House.

Jensen, O.B. (2007) City of layers. Bangkok's Sky Train and how it works in socially segregating mobility patterns. *Swiss Journal of Sociology* 33 (3), 387–405.

Jensen, O.B. (2009) Flows of meaning, cultures of movements – urban mobility as meaningful everyday life practice, *Mobilities* 4 (1), 139–158.

Jones, P.M. (1987) Mobility and the individual in western industrial society. In: Nijkamp, P. and Reichman, S. (eds.), *Transportation Planning in a Changing World*, 29–47. Aldershot: Gower.

Kaufman, V. (2002) *Re-Thinking Mobility*. Aldershot: Ashgate.

Knie, A. (1997) Eigenzeit und Eigenraum: Zur Dialektik von Mobilitat und Verkehr. *Soziale Welt* 47 (1), 39–54.

Langan, C. (2001) Mobility disability. *Public Culture* 13 (3), 459–484.

Lash, S. and Urry, J. (1994) *Economies of Signs and Space*. London: Sage.

Latour, B. (1999) *We Have Never Been Modern*. Cambridge, MA: Harvard University Press.

Löfgren, O. (2008) Motion and emotion: Learning to be a railway traveler. *Mobilities* 3, 331–351.

Lucas, K. (2009) *Scoping Study of Transport and Social Exclusion in the South African Context Report for the Republic of South Africa Department of Transport* (internal publication).

Lucas, K. (2011) Making the connections between transport disadvantage and the social exclusion of low income populations in the Tshwane Region of South Africa. *Journal of Transport Geography* 19 (6), 1320–1334.

Lucas, K. and Porter, G. (2016) Mobilities and livelihoods in urban development contexts: Introduction. *Journal of Transport Geography* 55, 129–131.

Maia, M.L., Lucas, K., Marinho, G., Santos, E. and Lima, J. (2016) Access to the Brazilian City – from the perspectives of low-income residents in Recife. *Journal of Transport Geography* 55, 132–141.

Mandri-Perrott, C. (2010) *Private Sector Participation in Light-Rail Metro Transit Initiatives*. Washington, DC: World Bank.

Massey, D. (1993) Power-geometry and a progressive sense of place. In: Bird, J., Curtis, B., Putnam, T., Robertson, G. and Tickner, L. (eds.), *Mapping the Futures: Local Cultures, Global Change*, 59–69. London: Routledge.

Massey, D. (2005) *For Space*. London: Sage.

Mlay, R., Chuwa, M. and Smith, H. (2010) Emergency referral transport needs of pregnant women in rural Tanzania. *African Journal of Midwifery and Women's Health* 4 (1), 5–13.

Morley, D. (2000) *Home Territories. Media, Mobility and Identity*. London: Routledge.

Mountz, A. (2011) Specters at the port of entry: Understanding state mobilities through an ontology of exclusion. *Mobilities* 6 (3), 317–334.

Mueller, F. (1994) Societal effect, organizational effect and globalization. *Organization Studies* 15 (3), 407–428.

Naumann, M. and Fischer-Tahir, A. (eds.) (2013) *Peripheralization: The Making of Spatial Dependencies and Social Injustice*. New York, NY: Springer Science and Business Media.

Nijkamp, P., Reichman, S. and Wegener, M. (eds.) (1990) *Euromobile: Transport, Communications and Mobility in Europe. A Cross-National Overview*. Aldershot: Averbury.

Ohnmacht, T., Maksim, H. and Bergman, M. (2009) Mobilities and inequality: Making connections. In: Ohnmacht, W.A., Maksim, H. and Bergman, M.M. (eds.), *Mobilities and Inequality*, 7–27. Aldershot: Ashgate.

Oviedo, H.D. and Dávila, J.D. (2016) Transport, urban development and the peripheral poor in Colombia – placing splintering urbanism in the context of transport networks. *Journal of Transport Geography* 51, 180–192.

Patton, J.W. (2004) *Transportation Worlds: Designing Infrastructures and Forms of Urban Life*, PhD thesis submitted to Rensselaer Polytechnic Institute. New York: Troy.

Porter, G. (2016) Reflections on co-investigation through peer research with young people and older people in sub-Saharan Africa. *Qualitative Research* 16 (3), 293–304.

Priya Uteng, T. (2011) *Gender-Related Issues Affecting Daily Mobility in the Developing Countries*. Washington, DC: The World Bank.

Priya Uteng, T. (2016) Transforming a new urban agenda into a just urban agenda (Based on presentation made by Prof. Adriana Allen). *PLAN* 6, 16–21.

Routledge, P. (2005) Grassrooting the imaginary: Acting within the convergence. *Ephemera* 5 (4), 615–628.

Scoones, I. (2009) Livelihoods perspectives and rural development. *The Journal of Peasant Studies* 36 (1), 171–196.

Secor, A.J. (2002) The veil and urban space in Istanbul: Women's dress, mobility and Islamic knowledge. *Gender, Place and Culture* 9 (1), 5–22.

Sheller, M. and Urry, J. (2006) The new mobilities paradigm. *Environment and Planning A* 38, 207–226.

Soja, E.W. (2000) *Postmetropolis: Critical Studies of Cities and Regions*. Oxford: Blackwell.

Sørensen, K.H. (1999) *Rush-Hour Blues or the Whistle of Freedom? Understanding Modern Mobility*, STS-working paper 3(99). Trondheim: Department of interdisciplinary studies of culture, Centre for technology and society.

Tiwari, G. (2003) Transport and land-use policies in Delhi. *Bulletin of the World Health Organization* 81 (6), 444–450.

UNFPA. (2007) *State of the World Population. Unleashing the Potential of Urban Growth*. New York, NY: United Nations Population Fund.

Urry, J. (2000) *Sociology Beyond Societies. Mobilities for the Twenty-First Century*. London: Routledge.

Urry, J. (2003) *Global Complexity*. Cambridge: Polity.

Urry, J. (2004) Connections. *Environment and Planning D: Society and Space* 22, 27–37.

Urry, J. (2007) *Mobilities*. Cambridge: Polity Press.

1 Cycling for social justice in democratizing contexts

Rethinking "sustainable" mobilities

Lake Sagaris and Anvita Arora

This chapter builds on field observation and experimentation in cities in developed and developing countries, and an ongoing collaboration between researchers in Chile, India, Canada, the United States and Europe. Our positionality combines experience with activism, planning, transport engineering and architecture, the fields in which we work.

Although urban transport systems are socio-technical hybrids, in practice they are seldom treated as such. Despite 50 years of contestation, transport engineers and planners have imposed a global system of "automobility" (Beckmann, 2001 ; Sheller and Urry, 2000) on cities in both the North and in the South. Perhaps the most evident result is the way that transport systems today reinforce exclusion, as highways segregate and isolate large communities of low-income families or slice through traditional neighbourhoods, expropriating current residents in favour of highways, subways and other transport infrastructure. Planning for the mobilities of a powerful minority traps the majority, generating multidimensional immobilities that threaten human, social and urban development.

This chapter explores a richer palette of definitions of "social" sustainability particularly as it relates to this (im)mobility paradox, highlighting the role of social agency for justice and equity as neglected, but crucial. We develop and use a framework to study how actions by "ecologies of actors" in specific niches of "transport mode ecologies" realize social sustainability through battles for social justice in transport policies, in Delhi (India) and Santiago (Chile).

In these experiences with transport-related conflicts, political agency emerges among grassroots organizations. As they interact with academic, governmental, public, other citizen and private actors, people challenge limited interpretations of economic or environmental sustainability, calling for new ways of co-creating sustainable mobilities for all, despite the global dominance of automobility.

Introduction: social sustainability and (im)mobilities in transport and cities

The editors of this book invite us to reflect on (im)mobilities in all the richness and tragedy of a term layered with meanings and contexts that, despite evident similarities, play out very differently in different contexts. From our perspective, as researchers exploring the interface between social imaginaries and needs, on the one hand, and the technical and logistical world of transport planning, mobility on the other hand. It involves the ability to move to better circumstances, food and safety, and is the difference between plants and animals. Immobility reflects a captivity that is undesirable and detrimental for human development and growth. Mobility is also the amount of movement one needs to make and is associated with distance, time and costs, which can become serious barriers to full participation in society, to growth and stability. In this chapter, we explore the paradoxes of (im)mobility in two urban settings: Delhi (India) and Santiago (Chile). While Delhi is witnessing ongoing battles of equity and inclusion, it has much to learn from the historical efforts of advocacy in Santiago which have led to a mature dialogue on social justice there. For this exploration, we start from "the complex patterning of people's varied and changing social activities" (Sheller and Urry, 2006: 213) specifically as they use, but also question, prevailing transport systems and the (im)mobilities they impose.

This chapter builds on field observation and experimentation in cities in developed and developing countries, and an ongoing collaboration between researchers in Chile, India, Canada, the US and Europe. Our positionality combines experience with activism, planning, transport engineering and architecture, the fields in which we work.

By nature, urban transport systems are socio-technical hybrids, but in practice they are seldom treated as such. Despite 50 years of contestation, transport engineers and planners have imposed a global system of "automobility" (Beckmann, 2001; Sheller and Urry, 2000) on cities north and south. Perhaps the most evident result is the way that transport systems today reinforce exclusion, as highways segregate and isolate large communities of low-income families or slice through traditional neighbourhoods, expropriating current residents in favour of highways, subways and other transport infrastructure. As one woman interviewed in Santiago said, "We finally get a Metro and a new express bus line on our road and they force us to leave! Who is this 'progress' for?"

Even in developed countries, particularly the United States where automobility originated, those without cars face serious limitations for full participation in society, whether through employment, education or other social benefits (Rosenbloom, 2004). This has led activists, practitioners and academics to focus on issues of "transport justice", while in the UK similar debates have emerged around "transport inclusion" (Lucas, 2004). There is less exploration of this intersection between a highly technified transport system and unjust and excluding social systems in developing countries.

Sometimes because of lack of contextual standards and sometimes because of aspirations, the design standards, cost-benefit analysis and other transport-planning methods from developed contexts mimic the sprawling suburban cities of the developed world in the Global South, imposing car-centred mobilities that reinforce and exacerbate inequalities and exclusion. In both Chile and India, the extreme concentration of wealth, privilege and power is reflected in high rates of car ownership among small segments of the population (Vasconcellos, 2001). The entire city must adjust to the needs of middle- and high-income drivers, sliced into isolated segments by major highways, with dangerous flyovers and interchanges, high levels of noise and air pollution.

In Chile, for example, imports of private cars have sky-rocketed in the past 30 years, but car ownership remains restricted to just 40% of households, some of which possess as many as five cars (SECTRA-UAHurtado, 2015). Both Chile and India have invested billions of dollars in highways and Metro lines serving mainly high- and middle-income neighbourhoods and users, while the public transport system, walking and cycling, which account for 73% of daily trips in Santiago and 72% in Delhi, are underfunded and underserved, exacting a high cost in human lives and serious disabilities. Thus, they create mobility traps rather than fluid mobilities that allow people to flow through the city, shifting modes according to needs, capacity, and motivation or trip purpose.

We explore these themes and how people respond in the context of two lively, dynamic urban centres in developing countries where these contradictions between (im)mobilities play out: Delhi and Santiago.

Transport: where social sustainability sparks action for justice

Transport is a critical link between economic and social development. Effective transport systems allow people to get to their jobs, take care of their health, pursue education and obtain the necessary food and goods to support their daily existence. For women, safe transport is essential for them to fulfil their roles in both the productive and the reproductive spheres, limiting or expanding their sphere of action according to which aspect – immobility or mobility – is favoured by the urban system overall. Likewise, poorly planned transport systems perpetuate inequities, increase air and noise pollution, raise barriers to access, segregate and ghettoize, compounding the complexity of urban planning dilemmas.

In the past decade, transport planning has sought more integrated and participative approaches to social development. Because transport strategies result from the complex interrelationships between the physical environment and social, economic and political activity, planning has become important to address community needs. This, however, cannot be accomplished without effective participation.

Stakeholder consultation, if not done with a democratic agenda, can bring in its own set of impositions. Decisions on investment can easily overlook needs and concerns of low-income and impoverished groups, especially in emerging economies, where resources are limited and projects compete in both the physical and social infrastructure sectors. Typical cost-benefit estimations of transport projects, for example, do not consider the costs imposed on cyclists and pedestrians, killed or permanently disabled by motor vehicles, when planning deliberately boosts speeds or road capacities.

Experience demonstrates that *broad-based* participation by affected *groups*/stakeholders in *decision-making* can ensure that transport improvements benefit the poor. Empowerment of local communities, through consultation, participation and ownership of local infrastructure, is also crucial for the social and financial sustainability of transport improvements (Gannon et al., 2001; Susskind and Elliott, 1983; Bickerstaff et al., 2002; Bickerstaff and Walker, 2005; Sagaris, 2014a; Sagaris and Landon, 2016).

The discussion illustrates that where policies are imposed by powerful governmental actors favouring the mobility of the few, conflicts result. These conflicts can shift perspectives on sustainability towards more socially inclusive mobilities.

The kinds of transport and social injustice discussed here reflect a disconnect between social welfare and the narrow economic approaches applied, for example, by multilateral development banks. These promise sustainability but continue to invest primarily in roads that serve only cars, rather than ensuring these serve and protect all users and modes, particularly walking, cycling and public transit. Elsewhere in the UN system, the United Nations Development Program (UNDP) notes that the relationship between development and economic growth is indirect, arguing that the real challenge is to transform economic growth into human development, paying particular attention to the redistributive mechanism necessary to ensure that any increase in income becomes increased prosperity for all (UNDP, 2007).

In this sense, and the Delhi case included in this chapter illustrates this well, employment is also an issue, or should be, when considering mobilities from a social justice perspective. Informal transport services such as jitneys, pedicabs and cycle rickshaws (and associated industries) employ large numbers of low-income people in many Asian cities, particularly in South Asia (Gallagher, 1992). Policies regarding these modes deeply affect the poor as customers, operators and employees. The issues involved may be complex. Much debate has occurred over what policies should be adopted foor the various 'non-corporate' transport modes, such as jitneys and pedicabs. However, from an anti-poverty perspective, reducing barriers to the informal supply of both passenger and goods transport is generally considered a "pro-poor" policy (Gannon and Liu, 1997).

These issues have catalyzed protests and social movements to contest transport injustice and achieve important innovations in sustainability and social inclusion. This is evident in the two cases explored here. The first occurred in Delhi when the government attempted to improve conditions for car drivers by severely restricting the use of traditional cycle rickshaws. Economic and efficiency arguments by powerful authorities were effectively contested by a grassroots coalition that mobilized support from academic, professional and other players. Similarly, in Santiago Chile, a major anti-highway movement in the city centre turned the spotlight on cycling as it could contribute to health, transport, access and new, more inclusive and fluid mobilities.

Cycling and walking are often treated as relevant but secondary elements in resolving sustainability challenges. Indeed, many World Bank investments seek to shift current car drivers into more "sustainable" modes, revealing a contradiction between environmental and social aspects of the sustainability paradigm. An environmental focus may lead funders to focus on expensive metros, bus rapid transit (BRT) systems or public bike share mainly in middle- and high-income communities, in an effort to persuade drivers to shift to more "sustainable" modes. This strategy deepens exclusion and segregation, by generating poor, disconnected walking and cycling facilities; depriving public transit of much needed investment to improve quality, comfort and safety; and neglecting the creation of low-cost mobility options, such as the cycle-rickshaws discussed later, which offer low-cost services, even as they provide significant income to needy households. Highways, moreover, further segregate low-income communities and generate high mortality rates as children and adults alike must cross them to reach schools, work and other vital destinations.

These examples suggest we need better definitions of the "social" aspects of sustainable mobilities, and how to operationalize them within social, spatial and transport planning. To achieve this, we have compared and contrasted our two cities to develop a framework focusing on a more precise definition of social sustainability. This has also required rethinking "sustainable" transport as an ecology of modes and users, interacting well or poorly within a given spatial and scalar environment.

Redefining and operationalizing social sustainability

Automobility followed closely on the heels of World War Two and largely coincided with what Huntington has called the "second wave" of democratization worldwide (1991). But with automobility came social mobilization and protest, and a democratization of urban planning in Europe, the US, Canada and Australia. Studies of these "anti-highway revolts", to use Ladd's label (2008), reveal a strong correlation between attempts by those in power to impose automobility, and local, regional and national movements to oppose it (Hovey, 1998; Hovey, 2003; Mohl, 2002; Mohl, 2004; Mohl, 2008; Fackler, 2009) and, most recently urban restoration and teardown movements by citizens and planning practitioners alike to restore the social aspects of cities (Mohl, 2012).

Writing about Portland (Thomson, 2001, Alarcon de Morris and Leistner, 2009), Vancouver (Harcourt et al., 2007), Toronto (Sewell, 1993) and New York (Jacobs, 1961; Flint, 2009; Sadik-Khan and Solomonow, 2016), diverse authors have revealed that a significant part of the socio-technical hybrid of urban transport systems involves active citizens, agency and transformative political action.

When applied to transport systems, however, "sustainability" tends to be interpreted mainly in terms of technical aspects such as energy and the environment, particularly air, water and noise pollution, with a significant focus on the economic impacts of oil dependency and congestion. Health stands out as the main social phenomenon associated with transport, through the growing focus by public health and urban planners on the "obesity epidemic" and the potential role for "active transport" (walking and cycling) in response.

A closer look at recent thinking about sustainable transport, reveals a rich set of principles that identify political agency, citizen participation in urban governance and a new set of social values as central to sustainability overall (Figure 1.1). Since the 1990s, several authors have argued for "smart growth" and

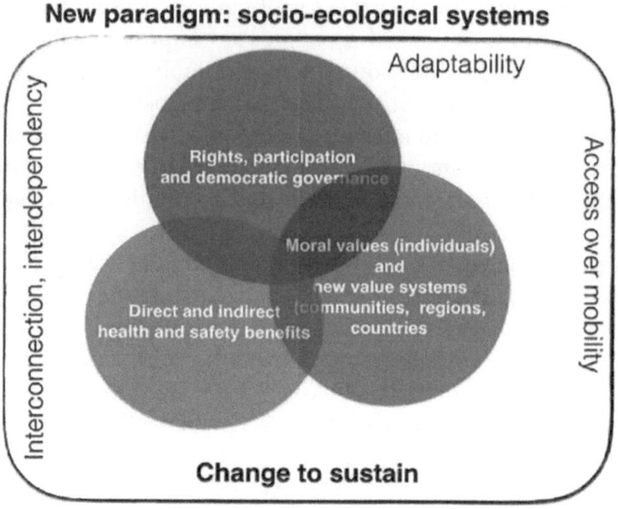

Figure 1.1 Key interactions relating to social sustainability, gleaned from authors' writings about social and governance issues as they relate to transport

Source: From figure 1, Sagris & Arora 2016.

better use of public transit (Cervero, 1998; Handy, 2005), while Low (2003) and other authors link transport closely to health (Frumkin, 2003; Frumkin et al., 2004). Low and Gleeson (2003) underline a significant paradox: the need to change in order to sustain.

Low and Gleeson (2003) emphasize the social imperative: to achieve sustainability, we must understand the importance of social institutions and that people "are capable of changing society and its institutions" (p. 3). Citing Polèse and Stren (2000), authors of one of the few book-length works on the social sustainability of cities, they centre social sustainability on justice and equity (p. 9). They argue for "eco-socialization", which "springs up within society's institutions and organizations as a new common sense (paradigm). . . . New questions are posed and answered. New values superimpose themselves over the old. . . . New rules and procedures are established to implement the new common sense. New technologies are gradually embedded to deliver the new values. Ecosocialization is a change at the level of the cells and organs that make up the body of society" (p. 19).

This perspective on social sustainability suggests that far from being equal, achieving environmental and economic sustainability of cities probably depends on achieving some significant, collectively recognized level of *social* sustainability. In recent years, issues of justice and equity have increasingly come to the fore in discussions of (social) sustainability. Lucas (2004) documents emerging strategies in the US and the UK to improve social justice and social inclusion, respectively as the basis for implementing social sustainability. To achieve these goals, requires a deeper theorization of social sustainability, one that goes beyond the typical interlocking circles of social development, economic growth and environmental protection, to consider how economic development is nested within society, which in turn is nested within an environment with important limitations. Indeed, Manzi et al. contextualize this nested approach further, arguing the need for a "multidimensional understanding" that underlines the importance of strengthening participation (the institutional imperative); safeguarding cohesion (the social imperative); improving competitiveness (the economic imperative); and recognizing limits (the environmental imperative). Within this multidimensional perspective, justice, democracy and eco-efficiency become central values, along with care, burden sharing and access (Figures 1.3 and 1.4; see also Figure 1.2, Manzi & Lucas, 2010).

Operationalizing this multidimensional approach, particularly in a specific field such as transport or mobilities, is both necessary and extremely difficult. One reason is the mobile nature of human identities, which are often treated as fixed variables that can be counted and quantified using criteria such as age, gender, religious or ethnic origins, etc. As Sheller notes, particularly in light of the social media so widely used today, individual identities are actually "a more-or-less

Figure 1.2 Key interactions in the sphere of social sustainability as they relate to transport are visible on Delhi streets, as pedestrians, cyclists, auto-rickshaws, motorcycles and other vehicles mix mobility and social activities.

Source: Sagaris field visit (2012).

Figure 1.3 Typical street scene in Delhi, India, reveals the unequal conditions for travellers, according to their mode, whether walking, cycling or using diverse variations on motorized modes

Figure 1.4 Points where walking, cycling and rickshaws can improve public transport service and safety for women and create jobs

rickety ensemble", part of mobile publics in a social world of "disorderly 'gels' and 'goos'". "Publics" become "special moments or spaces of social opening that allow actors to switch from one setting to another, and slip from one kind of temporal focus to another" (Sheller, 2004: 48).

This view of the social as constantly changing, with fuzzy borders and virtually undefinable characteristics, underlines one of the most powerful features of the wicked problem of sustainable transport: people themselves. If, as Sheller argues, people cannot be defined solely by age, socio-economic or other demographic characteristics, how do we study and influence their relationship to the transport system? Roles shift constantly, from one set of relations to another, and every actor carries "multiple identifications and capacities to 'play' different parts at once" (Sheller, 2004: 48).

Banister (2005, 2007, 2008), alone or with other authors (Banister et al., 2007; Banister et al., 2012) concurs, offering a particularly rich view of the *political* dimensions of social sustainability and transport. To realize the values mentioned previously, Banister argues for community participation and strong civil society. "Moral capital", based on diversity and rights, should complement investment in human and social capital.

Cox (2010) too highlights the political, particularly *agency*, using Delhi to demonstrate how grassroots campaigning takes participation beyond the limits of government-controlled consultations, characterized as paternalism by Susskind and Elliott (1983). (Re)education is necessary, as we see with advocacy and Safe Routes to School programs. Public bike share demonstrates cycling's usefulness for improving public transport. Innovative projects in Uganda, South Africa and Namibia, often led by local actors rather than governments, reveal potential for job and income generation of cycling-related businesses. In Latin America, tricycles remain a vital form of transport for cargo and for gardeners, plumbers and other independent trades people, although many transport planners do not even realize they exist. In South Asia, bicycle taxis and rickshaws yield social and other benefits.

Cahill (2010) notes the importance of social *movements*, whether against highways or campaigning for sustainable transport. He explores health, gender (related to cars), the loss of spaces for children's play, and issues for older people, arguing for directly linking transport and social policy. Like the authors discussed here, he defines sustainability in terms of modes, but also citizenship and social justice.

Figure 1.1 summarizes key interactions regarding social sustainability, as compiled from a review of these and other authors. We have put social agency – rights, participation and democratization of urban and regional governance – at the top, as this indicates they are paramount to generating paradigmatic change. People contest the dominance of automobility through specific conflicts, and through those conflicts they expand thinking about how transport systems should be planned, funded and regulated. These conflicts also catalyze changes in individual attitudes and collective behaviour: how the system can be used to create new and more equally shared mobilities.

Agency is not the purview of a single type of actor, although organized citizens, that is, civil society often take the lead. "Inside activists" within government and the private sectors amplify demands and implement changes proposed by citizens (Olsson and Hysing, 2011), underlining the role of *interactions* among visionary political leaders and organized citizens, an "ecology of actors" (Evans, 2002).

An "ecology of modes" to include walkability and cycle-inclusion

Based largely on our observations and conversations in Delhi (Figure 1.3, 2012), we began to think about automobility as a kind of "monomode", like the monocrops criticized in much of the literature on agriculture. This led to our considering "sustainable transport" as an ecology of modes, in which the interactions between environmentally friendly, energy-efficient modes, including those based on human and animal energy, were crucial to the overall sustainability of available mobilities.

Delhi has a particularly rich ecology of modes, ranging from walking and cycling, including cycle taxis (rickshaws) used for both passengers and cargo, through auto rickshaws, motorcycles, cars, buses and Metro (Table 1.2). An Ashoka fellow, Pradip Kumar, has demonstrated how redesigning a better rickshaw, and innovation in business models can improve income distribution and make good use of smart phone applications. This approach can significantly improve the flow of passengers between public transport and the "last mile", which is a barrier to families with small children, women and others with large burdens, and those with physical disabilities (Shanbaug, 2012).

This "intermodal" approach focuses on the interactions between diverse modes, whose niches can be defined according to the distance covered in a reasonable period (Table 1.1). Together with land use policies that focus on trip purposes

Table 1.1 Optimal distances and times for walking and cycling as standalone and public transport access modes

	Walk (minutes)			*Cycle (minutes)*		
m	*Easy (4.5 km/h)*	*Moderate (5 km/h)*	*Fast (5.5 km/h)*	*Easy (15 km/h)*	*Moderate (19 km/h)*	*Fast (24 km/h)*
400	5.3	4.8	4.4	1.6	1.3	1
800	11	9.6	8.7	3.2	2.5	2
1,200	16	14	13	4.8	3.8	3
2,000	27	24	22	9	6.3	5
3,000	40	36	33	12	9.5	7.5
4,000	53	48	44	16	13	10
5,000	67	60	55	20	16	13
8,000	107	96	87	32	25	20
10,000	133	120	109	40	32	25

Source: Table 1.1, Karner and Sagaris, 2016. Note: white = reasonable time for travel to access public transport; light grey = standalone single-mode travel or cycle service trip (bike taxi, bike share, etc.); dark grey = best served by combination with motorized modes.

Table 1.2 Modal shares in Delhi

Mode	*Avg speed** km/h*	*Avg time min.*	*Avg trip length km*	*Trips under 5 km %*	*Trips over 45 minutes %*	*Modal share* %*	*Trips by mode n*	*Occupancy factor*
Walk	4	10	1	100	0	11	1,879,500	1
Cycle	13	25	5	75	14	26	4,522,000	1
Motorized two-wheeler	20	27	9	21.74**	29**	15	2,576,000	1
Auto rickshaw	20	29	10	45	13	1	134,750	2
Car	20	32	11	12.87**	31.2**	13	2,189,250	1
Bus	18	49	15	15	50	35	6,198,500	56
Total						100	17,500,000	

Source: Table 1.2, Arora, 2013, based on RITES, 2005; RITES 2001; TRIPP, 2000 (*); Mukti, 2009 (**).

other than work, such as education, accessing health care or culture, care for others and shopping, intermodal planning would permit establishing modal share *targets*, reassigning short motorized trips to walking (0–2 km) and diverse formats of cycling (2–8 km), with public transport prioritized in the medium- to high-density urban areas that characterize most cities, and travel by car favoured mainly for long-distance journeys in low-density areas, that is, suburban and interurban trips.

To test the usefulness of our observations from Delhi, we conducted an analysis of both San Francisco, a car-intensive city, and Santiago, a transitioning city with soaring car use. The analysis revealed that between 50 to 75% of car trips are under 8 km, thus candidates for shifting to more sustainable modes (Karner and Sagaris, 2016; Sagaris and Arora, 2017). This approach could generate a new set of mobilities based on human energy and policies that could exclude or minimize private car use in high-density residential and destination areas, as Barcelona is beginning to do with its "super-block" approach (www.theguardian.com/cities/2016/may/17/superblocks-rescue-barcelona-spain-plan-give-streets-back-residents). From this perspective, mobilities based on human energy become central to resolving major challenges of social equity, inclusion or global warming (Sagaris and Arora, 2017).

How can city governments hope to convince citizens that radically reducing car use, or major investment in segregated bus- and cycle-ways is beneficial not only to individual users, but also to the city as a whole? Focusing solely on one mode for the most privileged sectors of society generates mobilities for a few, at the expense of producing enormous immobilities for the majority. Our analysis of the conflicts arising over cycling – as a private transport mode (Santiago) and as an economic activity (Delhi) – indicates that these urban (im)mobilities are important sites of conflict, offering significant opportunities for innovation.

Effectively challenging automobility requires deliberative participation, which can be catalyzed by conflicts led by independent citizen movements and "deep" processes that bring citizens into direct contact with decision-makers. These instances of participation and the civil society organizations that result should be nourished and supported as an integral part of the planning system. Otherwise, the opportunity they represent may be under-realized.

In this chapter, we use this framework combining social sustainability as operationalized through action by ecologies of actors in specific niches of transport mode ecologies to better understand and extract lessons for socially just and inclusive sustainability in Delhi and Santiago. In these transport-related conflicts, we see political agency emerge. People in very different contexts learn similar skills, interacting expertly with academic, governmental, public, other citizen and private actors, challenging limited interpretations of economic or environmental sustainability, and calling for new ways of co-creating sustainable mobilities for all, despite the global dominance of automobility.

Ecologies of modes and actors in Delhi: the battle for cycle rickshaws' survival

The National Capital Territory of Delhi is India's largest metropolitan region by area. Located between major shopping areas, it is surrounded by industrial zones, all key employment areas. Unlike most Indian cities, Delhi has a lower modal share for cycling (see Table 1.2). Nonetheless it remains important, with up to 4.5 million commuter trips daily. Cycle traffic accounts for more than 19% of traffic in peak hours and involves primarily short trips, whether by individuals or as part of services in low-, middle- and high-income areas. Notwithstanding, 14.4% of cycle trips are longer than 45 minutes, an indication that longer trips are viable and relevant. An average cycle ride takes 25 minutes to cover 5.5 km, with 75% of trips below 5 km in length, and total modal share 26%, despite the lack of specialized infrastructure or measures to preserve cyclists' safety (see Table 1.2, Arora, 2013).

The cycle rickshaw was introduced in Delhi in 1940 and today an estimated 600,000 (Manushi, 2010) cycle rickshaws ply the roads of Delhi, travelling 25–30 km/day as they carry six to eight passenger trips, or a total of 1.5 million trips per day. These sustain more than 2 million people, despite limitations imposed by city authorities to improve flows of motorized vehicles, mainly cars. Rickshaws provide access, mainly to middle- and high-income women, accounting for 23% of ingress/egress trips to public transport. Figure 1.4 illustrates the points in the intermodal trip typically covered by cycle rickshaws (Kumar and Parida, 2011).

A study by Jan Parivahan Panchayat ("People's Transport Council," a working group of Lokayan, a movement for development alternatives) notes that rickshaw pulling provides a livelihood for seasonal migrants (66% of rickshaw pullers) from impoverished villages and generates employment opportunities to pullers, mechanics and others in the small-scale industries producing rickshaw parts (Nandhi, 2011). The rickshaw provides not only last-mile connectivity from the bus and metro but also covers short trips to markets and schools and goods delivery in crowded market places.

The Delhi government has a history of marginalizing this mode, however. In a city that has no limits on the number of cars that can be bought and run, the Municipal Corporation of Delhi (MCD, the local governing body) capped the maximum number of licensed rickshaws in the city – one person is granted one cycle rickshaw license, except widows and handicapped persons, who can hold five (CPPR, 2009). Plying a cycle rickshaw without a license is illegal. The initial cap on cycle rickshaws, put in place in 1960, was 750.

From time to time, the cap was revised, but never considered actual needs. In 1975, the MCD increased the number to 20,000. Eighteen years later, it raised the cap to 50,000, despite actual numbers standing around 450,000. Four years later, the MCD raised the cap to 99,000, forcing most of the 600,000 cycle rickshaws to remain outlaws (Manushi, 2010). To compound this misery, city police divided Delhi into Red, Blue and Yellow zones. In the Red Zone, all rickshaws were banned, while in the Blue Zone, they could operate 24 hours/day. In the Yellow Zone, they were banned from 8am to 11pm. The Red and Yellow zones overlapped, so it was difficult to know where to operate. Passengers, moreover, expect rickshaws to travel across different zones. If a rickshaw driver was caught, traffic police would drive nails into tyres, demand bribes, damage or confiscate vehicles.

The gap in permits and the actual numbers plying and the limits on area of operation outlawed the vast majority of rickshaw pullers, worsening their vulnerability. MCD officers impounded and dismantled the confiscated rickshaws (Figure 1.5) if the owner or puller failed to pay the fine and storage fees. In practice,

Figure 1.5 Cycle rickshaws being confiscated in Delhi

pullers and owners often had to pay a bribe to get their rickshaws back. According to Manushi (2001), the bribes alone totalled approximately US$1.5 million/month, making this situation profitable and desirable for those in power.

The All Delhi Cycle Rickshaw Operators' Union, Federation of Rickshaw Pullers of India (FoRPI), Rajdhani Cycle Rickshaw Pullers' Union and many small rickshaw pullers' unions joined with major trade unions and NGOs to organize rickshaw pullers and bring them under an umbrella union to safeguard their right to livelihood and recognition for their role in society. Unions have been trying to coordinate between city planners, policymakers, police and municipal bodies to get rickshaw pullers proper identification and social securities. Unions and NGOs have also been working towards providing help lines and basic necessities such as night shelters, warm clothes in the winter, parking rights and medical support in case of accidents or serious illness. Alcoholism and drug use are common among rickshaw pullers. Several civil society groups are also working to create awareness among pullers about the ill effects of alcoholism and drug abuse. Providing financial and social security to all rickshaw pullers has been another priority of unions.

Several unions filed lawsuits against the unjust regulations. As early as 1987, the All Delhi Cycle Rickshaw Operators Union petitioned the Delhi High Court, arguing the MCD by-law breached Article 19(1)(g) of the Constitution of India, which guarantees the right for a citizen to practice any profession, occupation, trade or business. Manushi (a welfare organization founded by social activist Madhu Kishwar) filed a case in the high court in 2007 questioning the government's policy on cycle rickshaws (Manushi, 2010). In 2010, the court ruled in favour of the rickshaw community, a significant victory, and a committee formed, including urban planners, transport planners, NDMC, police, etc. The committee was assigned the task of producing an alternative policy that was rickshaw inclusive.

The Delhi government challenged this order in the Supreme Court, which ruled in favour of rickshaws. The court argued that putting a cap on cycle rickshaws alone in Delhi is unjust and the MCD's act directly denied rickshaw pullers the right to decent work, earn a livelihood and live with dignity (Venkatesan, 2010). The court further argued that constitutional rights should be interpreted in context and took serious note of the pathetic conditions in which the city's migrants lived. The court questioned the MCD's motives for applying such controversial rules.

Both media and urban planning professionals eventually supported the rickshaw pullers, which allowed for the lifting of the various bans against the rickshaw. Since then, the city has included the rickshaws in infrastructure planning and the new Masterplan of Delhi explicitly recognizes their role in meeting the city's mobility needs.

Transport planners and civic authorities often do not understand the livelihood implications of transport system–related decisions – in this case banning the cycle rickshaws ostensibly to counteract congestion and the growing demand for space by the automobiles. This conflict illustrates how transport has a critical social role in creating jobs and opportunities for owners, pullers and users, thereby generating new, sustainable, more inclusive mobilities that consider employment as well as service.

Lessons from activism's shift to effective advocacy in Santiago, a transitioning city

From 1990 onward, Chilean society began to emerge from the terror and violence that dominated all social interchange during the 17 years of military government (1973–1990). Governance, however, remained extremely authoritarian, with the national constitution, government and major institutions still shaped by the constitution developed in the 1980s, on the assumption that the regime could last virtually forever. One of the first laws generated by the new democracy reflected the emergence, during the 1970s and 1980s, of environmental concerns. Although imperfect, this new environmental law and the institutions it created began to generate significant spaces for conflict, which led to considerable success in protecting wilderness areas in Tierra del Fuego, for example, or fishing villages in southern Chile.

In the late 1990s, the country experienced its first major environmental conflict in an urban, rather than a wilderness or rural context: 25 diverse community organizations united to oppose a major highway concession, the Costanera Norte. The project was the brainchild of then public works minister, Ricardo Lagos, a charismatic leader of the Socialist Party, on his way to the national presidency. Twenty-three kilometres in length, it was also the first urban highway concession project, that is, major infrastructure ostensibly built and run by private companies in partnership with the national government. (In practice, it required more than US$500 million in direct subsidies, sapping resources from much-needed public transit.)

The project was designed to run from the wealthy eastern suburbs to the airport and the coast to Santiago's west, running straight through the middle of the city, with its traditional market district, the Bellavista arts neighbourhood, and a mix of low-, middle- and high-income areas. It was these central, highly visible and well-connected communities that overcame the legacy of fear and paralysis left by the military regime. With support from environmental and other civil society organizations, neighbourhood and market vendors' associations were able to combine forces, creating a coalition *Coordinadora No a la Costanera Norte*.

Characterized as "deliberate improvisation" (Silva, 2011), the process that ensued matched an all-powerful public works ministry against seemingly small, powerless neighbourhood groups, with little funding or knowledge about urban and transport planning. Notwithstanding, they successfully opposed the project, building strong, independent citizenship among leaders and supporters, at the same time as they raised important questions about democracy and urban governance (Ducci, 2000; Ducci, 2002; Sagaris, 2010; Sagaris, 2014b).

Although the highway was eventually built, it runs under the river in the Coordinadora's territory, a decision that saved three of its four main neighbourhoods from destruction. Leaders then founded Living City (2000), a citizen-led planning organization that has since run several award-winning participatory projects, involving neighbourhood recovery, safety and security (against drug trafficking networks and other forms of urban violence), recycling and sustainable transport. Most noteworthy was a Citizen-Government Roundtable, developed by Living City with partners in four other pro-cycling organizations. These united to create a United Cyclists of Chile network (CUCH) in 2006, to move beyond a critical mass-inspired monthly cycle ride that had been the sole expression of cycling advocacy from 1997 on. In the early 2000s, the machismo of male leaders of the cycling movement led to the founding of Macletas, a women's cycling group; Bicicultura, a mixed pro-cycling initiative that focused on cultural and political means of change; and CicloRecreovía, the Chilean version of the Latin American practice of closing streets to cars on Sunday, which began in Bogotá and has now spread worldwide.

CUCH operated on rules similar to those developed by the anti-highway coalition. Each group retained its leadership and its spokepersonship, collaborating with all or some other groups, depending on the issues at hand. Both the Bicicultura festival and a Citizen-Government Roundtable for cycle inclusion were central activities that together had a major impact on cycle use in metropolitan Santiago.

Despite many setbacks (Sagaris and Ortuzar, 2015), the roundtable functioned for three years and produced Santiago's first major cycle plan with public involvement (Sagaris and Olivo, 2010). The process was strongly influenced by support from a Dutch NGO, Interface for Cycling Expertise, which invested significantly in civil society, providing intensive training for citizens and planners alike, as they strove to improve cycle inclusion. By the end of the three-year process, cycling infrastructure had quadrupled to 200 km, cycle parking had improved significantly (both design and locations), a women's cycling school had trained more than 1,000 new riders, and the city was in the midst of a cycling revolution. Indeed, cycling's image went from "a poor man's ride" (exemplified by a popular advertisement to this effect) to being the trendy way to get around, with modal share doubling from 2% to 4%, as revealed by the 2012 origin-destination survey. Counts, which tend to be more accurate, show that the number of cyclists on the main cycle ways is rising by 25 to 30%; a public bike share exists in 16 of the region's 34 municipal *comunas*; women cyclists account for up to 30% of riders (up from 12% in 2004, when the Macletas began). New regulations and standards are helping to improve the rather appalling quality of the first cycleways, and pioneering local governments have begun to integrate walking-cycling-public transport improvements, which are reducing the space available to cars.

A detailed review of the growing body of science regarding cycle planning and its effects (Sagaris, 2015) identified the importance of interactions between measures in the urban sphere (road speeds, infrastructure, laws), the social sphere (education for behavioural change, campaigns to foster more awareness of cycling and its benefits) and the economic sphere (jobs, businesses and services necessary for cycling to flourish and attract support from diverse people). This echoes Sheller and Urry's insistence on moving beyond "bipolar" readings of complex realities:

debates of infrastructure versus education abound, but the evidence clearly shows that approaches combining both, and the economic elements described here, are crucial.

In Chile at least, these lessons are seldom applied in transport planning, where the idea that transport planning is primarily a "technical" question continues to prevail. Notwithstanding, even in Chile, there is strong evidence that lack of social involvement and particularly empowered participation to improve both walking and public transport, is seriously hampering efforts to achieve more sustainable living.

Final reflections: implications for planning and further research

As the experiences discussed here reveal, issues of agency raised by Sheller and Urry (2006) have important political implications as diverse social movements and organizations contest the (im)mobilities imposed by the globalization of car-centred urban planning, automobility. In developing countries such as Chile and India, imposition reinforces and deepens existing exclusions and injustices, by reducing both individual mobility and crucial economic activities that serve diverse users at moderate prices.

These experiences suggest that non-motorized, or active, transport will play a crucial role in twenty-first century mobilities systems, even where they are "organized around new 'machines' enabling 'people' to be more individually mobile through space, forming small world connections 'on the go'. 'Persons' will occur as various nodes in multiple machines of inhabitation and mobility". They exemplify the existence of human-energy propelled "machines inhabited by very small groups of individuals" (Sheller and Urry, 2006: 221), enacting vital connections within ecologies of social actors and transport modes.

These conflicts and their achievements also underline the importance of the local, disturbing the "bipolar logic" that focuses on the separations rather than the links between scales. Law (2004) asks what "if the global were small and noncoherent?", contrasting the traditional "romantic view" of its overweening importance with a "baroque" alternative, that considers the "limitless internal complexity" in the local. This is consistent with complexity thinking: based in ecology (Gunderson and Holling, 2002), it offers the concept of "nested scales", interacting within and among each other, simultaneously. In these cases, our local, apparently "small" groups of actors organized effectively to challenge whole governance and urban management regimes. They did so to defend neighbourhoods against highways and favour cycle inclusion (Chile), or to save cycle-related livelihoods (Delhi).

Thus, bicycles and tricycles, typically considered "small", not very relevant players in the search for urban transport solutions, have catalyzed a level of agency that is sorely needed by, for example, diverse public transport schemes, particularly those based on buses. Sheller and Urry's "new" machines may evolve from old ones, cast out in contempt by automobility, but required for new mobilities.

Software may increasingly "write mobility" (Sheller and Urry: 221), in the sense that specific software systems "need to speak effectively to each other in order that particular mobilities take place". That mobility, however, could well involve preserving and building on high modal shares for the walk-bike-bus combinations that still account for the vast majority of trips in developing countries, even after years of automobility. New technologies, sometimes associated with "Smart" cities, such as systems for counting people walking and cycling, applications to improve women's safety or to ensure a cycle taxi is waiting at the precise time that a passenger and her children arrive at a particular bus stop, can improve the interface between these sustainable modes, and thereby enhance their social benefits, in terms of inclusion, respect and income. But there is a clear risk that they will simply be overrun, if technology reinforces a neoliberal focus on the large over the small, operator, business or customer.

These experiences underline the importance of understanding "sustainable transport" as a socio-technical hybrid, at different scales and layers of intensity. Like cars, buses and metro systems are (to date) both human and technological systems. Cycling combines both too. When skilfully combined, a new system emerges that constitutes "sustainable" transport. This requires going beyond planning of individual modes, or even multi-modality, to focus on the *inter*-modal. An intermodal focus, then, makes interconnections among modes, people and contexts (social spaces, moving places and land use) as much a part of these new mobilities as the human-natural-artefacts moving through them.

Our research on cycle inclusion and intermodality suggests that, rather than long lists of measures, their success depends primarily on how human agency, land use and transport planning interact to generate new mobilities (Sagaris, 2015; Sagaris and Arora, 2015). As these cases illustrate, the crucial factor is the mobilization of ecologies of actors, particularly *organized* citizens, who can challenge discriminating and excluding systems. Thus, the most important investment to achieve more sustainability mobilities is to complement infrastructure and technology, with the social – and civic – components able to catalyze major, paradigmatic change.

Acknowledgements

This research was supported by Canada's Social Science and Humanities Research Council and the Centro de Estudios de Desarrollo Urbano Sustentable (CEDEUS), with funding from Conicyt, FONDAP No. 15110020. We would also like to thank reviewers and colleagues in the Centre for Excellence in BRT (Pontificia Universidad Católica de Chile), and community leaders who shared their experiences with us, from the Institute for Democracy and Sustainability (Delhi) and Ciudad Viva/Living City (Chile).

References

Alarcon de Morris, A. and Leistner, P. (2009). From neighborhood association system to participatory democracy broadening and deepening public involvement in Portland, Oregon. *National Civic Review* Summer: 48–55.

Arora, A. (2013) *Non-Motorized Transport in Peri-urban Areas of Delhi, India*. Case study prepared for Global Report on Human Settlements 2013.

Banister, D. (2005) *Unsustainable Transport: City Transport in the New Century*. London: New York, NY: Routledge.

Banister, D. (2007) Sustainable transport: Challenges and opportunities. *Transportmetrica* 3, 91–106.

Banister, D. (2008) The sustainable mobility paradigm. *Transport Policy* 15, 73–80.

Banister, D., Pucher, J. and Lee-Gosselin, M. (2007) Making sustainable transport politically and publicly acceptable: Lessons from the EU, USA and Canada. *Institutions and Sustainable Transport: Regulatory Reform in Advanced Economies*, 17–50. Cheltenham, UK: Edward Elgar Publishing.

Banister, D., Schwanen, T. and Anable, J. (2012) Introduction to the special section on theoretical perspectives on climate change mitigation in transport. *Journal of Transport Geography* 24, 467–470.

Beckmann, J. (2001). Automobility a social problem and theoretical concept." *Environment and Planning D: Society and Space* 19, 593–607.

Bickerstaff, K., Tolley, R. and Walker, G. (2002) Transport planning and participation: The rhetoric and realities of public involvement. *Journal of Transport Geography* 10, 61–73.

Bickerstaff, K. and Walker, G. (2005) Shared visions, unholy alliances: Power, governance and deliberative processes in local transport planning. *Urban Studies* 42, 2123–2144.

Cahill, M. (2010) *Transport, Environment and Society*, Maidenhead, UK: McGraw-Hill Education.

Cervero, R. (1998) *The Transit Metropolis : A Global Inquiry*. Island Press; Washington, DC.

Cox, P. (2010) *Moving People : Sustainable Transport Development*. London, New York, NY, Cape Town, South Africa: Zed; UCT Press; Distributed in the USA by Palgrave Macmillan.

CPPR. (2009) *MCD Cycle Rickshaw Rules and Regulations*. Available at: www.livelihood-freedom.in/L%203%20FOR%20SITE/Delhi/CYCLE%20RICKSHAW.pdf.

Ducci, M.E. (2000) *Governance, urban environment, and the growing role of civil society*. Project on Urbanization, Population, Environment and Security. Washington, DC: Woodrow Wilson International Centre for Scholars, 18.

Ducci, M.E. (2002) The importance of participatory planning, the principal Urban struggles of the New Millennium. In: Ruble, BA., Stren, R., Tulchin, J.S., et al. (eds.), *Urban Governance Around the World*, 153–186. Washington: Woodrow Wilson International Centre for Scholars

Evans, P.B. (2002) *Livable Cities? Urban Struggles for Livelihood and Sustainability*. Berkeley: University of California Press.

Fackler, E.H. (2009) *Protesting Portland's Freeways: Highland Engineering and Citizen Activism in the Interstate Era*. University of Oregon, Department of History, Eugene Oregon, 133.

Fackler, E. H. (2009) *Protesting Portland's Freeways: Highland Engineering and Citizen Activism in the Interstate Era*. M.A. Master of Arts, University of Oregon, Department of History.

Flint, A. (2009) *Wrestling with Moses How Jane Jacobs Took on New York's Master Builder and Transformed the American City*. New York, NY: Random House.

Frumkin, H. (2003) Healthy places: Exploring the evidence. *American Journal of Public Health* 93, 1451–1456.

Frumkin, H., Frank, L.D. and Jackson, R. (2004) *Urban Sprawl and Public Health: Designing, Planning, and Building for Healthy Communities*. Washington, DC: Island Press.

Gallagher, R. (1992) *The Rickshaws of Bangladesh*. Dhaka: University Press.

Gannon, C.A. and Liu, Z. (1997) *Poverty and Transport*. Discussion Paper, TWU Papers TWU-30. Transportation, Water and Urban Development Department (TWU), World Bank, Washington, DC.

Gannon, C.A., Liu, Z., and Calvo, Malmberg C., (2001) Transport: Infrastructure and Services, Draft for comments, World Bank, Washington DC.

Gunderson, L.H. and Holling, C.S. (2002) *Panarchy: Understanding Transformations in Human and Natural Systems*. Washington: Island Press.

Handy, S. (2005) Smart growth and the transportation-land use connections: What does the research tell us. *International Regional Science Review* 28, 146–167.

Harcourt, M., Rossiter, S. and Cameron, K. (2007) *City Making in Paradise: Nine Decisions That Saved Vancouver*. Vancouver: Douglas McIntyre.

Hovey, B. (1998) Building the city, structuring change: Portland's implicit utopian project. *Utopian Studies* 9. 68–79.

Hovey, B. (2003) Making the Portland way of planning: The structural power of language. *Journal of Planning History* 2, 140–174.

Huntington, S.P. (1991) *The Third Wave: Democratization in the Late Twentieth Century*. Norman: University of Oklahoma Press.

Jacobs, J. (1961). *The Death and Life of Great American Cities*. Vintage Book, New York. 458 p.

Karner, A. and Sagaris, L. (2016) *Testing a new approach to planning sustainable transport using data from metropolitan Santiago de Chile and the San Francisco Bay area*. Transportation Research Board. Washington, DC: Transportation Research Record.

Kumar, P. and Parida, M. (2011). Vulnerable road users in multi modal transport system for Delhi. *Journeys* May, 38–47.

Ladd, B. (2008) *Autophobia: Love and Hate in the Automotive Age*. Chicago: University of Chicago Press.

Law, J. (2004) And if the global were small and noncoherent? Method, complexity, and the baroque. *Environment and Planning D: Society and Space* 22, 13–26.

Low, N. (2003) The active city. *Urban Policy and Research* 21, 5–7.

Low, N. and Gleeson, B. (2003) *Making Urban Transport Sustainable*. Houndmills, Basingstoke, Hampshire, New York, NY: Palgrave Macmillan.

Lucas, K. (2004) *Running on Empty: Transport, Social Exclusion and Environmental Justice*. Bristol, UK: Policy.

Manushi. (2001) *Wheels of Misfortune: The License-Quota-Raid Raj and Rickshaw Pullers*. Available at: www.indiatogether.org/manushi/rickshaw/.

Manushi. (2010) *Public Interest Litigation: Manushi vs. Government of Delhi*. Available at: www.manushi.in/tinymce/uploaded/Rickshaw%20Judgement-High%20court.pdf.

Manzi, T. and K. Lucas (2010) *Social Sustainability in Urban Areas*. New York, Earthscan.

Mohl, R. (2002) *The Interstates and the Cities: Highways, Housing, and the Freeway Revolt*. Birmingham, AL: Department of History (University of Alabama at Birmingham), Poverty and Race Research Action Council.

Mohl, R. (2004) Stop the road: Freeway revolts in American cities. *Journal of Urban History* 30, 674–706.

Mohl, R. (2008) The interstates and the cities: The US department of transportation and the freeway revolt 1966–1973. *Journal of Policy History* 2, 193–226.

Mohl, R.A. (2012) The expressway teardown movement in American cities. *Journal of Planning History* 11, 89–103.

Nandhi, A.M. (2011) *The Urban Poor and Their Money: A Study of Cycle Rickshaw Pullers in Delhi.*

Nandhi, M. A. (2011). *The Urban Poor and Their Money: A Study of Cycle Rickshaw Pullers in Delhi.* Delhi, India: Microfinance Researchers Alliance Programa, Centre for Microfinance, University of Delhi.

Olsson, J. and Hysing, E. (2011) Theorizing inside activism: Understanding policymaking and policy change from below. *Planning Theory Practice* 13, 1–17.

Polése, M. and Stren, R. (2000) *The Social Sustainability of Cities: Diversity and Management of Change.* Toronto: University of Toronto Press.

Rosenbloom, S.C. (2004) Research on women's issues in transportation. In: Board, T.R. (ed.), *Research on Women's Issues in Transportation.* 41–49, 59–67. Chicago, IL: Transport Research Board.

Sadik-Khan, J. and Solomonow, S. (2016) *Streetfight: Handbook for an Urban Revolution.* New York, NY: Viking.

Sagaris, L. (2010) Learning democratic citizenship neighbourhoods as key places for practicing participatory democracy. In: Pinnington, E. and Schugurensky, D. (eds.), *Citizenship Learning and Participatory Democracy Throughout the World.* Cambridge Scholars Publishing, Newcastle upon Tyne, UK.

Sagaris, L. (2014a). Citizen participation for sustainable transport: The case of 'Living City' in Santiago, Chile (1997–2012) *Journal of Transport Geography* 41, 74–83.

Sagaris, L. (2014b) Citizens' anti-highway revolt in Pinochet Chile: Catalyzing innovation in transport planning. *Planning Practice and Research* 29, 268–280.

Sagaris, L. (2015) Lessons from 40 years of planning for cycle-inclusion: Reflections from Santiago, Chile. *Natural Resources Forum* 39, 64–81.

Sagaris, L. and A. Arora (2016). Rethinking "Sustainable" Transportation as Bike-Bus Intermodal Integration. In Henscher, D. and Muñoz, J. C.*Thredbo International Series.* 59, 218–227. Santiago, Chile.

Sagaris, L. and P. Landon (2017) Autopistas, ciudadanía y democratización: la Costanera Norte y el Acceso Sur, Santiago de Chile (1997–2007). *EURE* 43 (128), 127–151.

Sagaris, L. and Olivo, H. (2010) *Plan Maestro de Ciclo Rutas del Bicentenario.* Santiago, Chile: Santiago Regional Metropolitan Government, Interface for Cycling Expertise, Living City, 60

Sagaris, L. and J. d. D. Ortuzar (2015). Reflections on citizen-technical dialogue as part of cycling-inclusive planning in Santiago, Chile. *Research in Transporation Economics.* **53** (November), 20–30.

SECTRA-UAHurtado. (2015) *Estudio de origen – destino santiago RM. Origin-Destination Surveys.* Santiago, Chile: SECTRA, Ministry of Transport, Government of Chile.

Sewell, J. (1993) *The Shape of the City: Toronto Struggles With Modern Planning.* Toronto: University of Toronto Press.

Shanbaug, A. (2012) 'Rickshaw Bank' concept changes lives of thousands of pullers. *The Economic Times.* New Delhi, India: India Times.

Sheller, M. (2004) Mobile publics: Beyond the network perspective. *Environment and Planning D: Society and Space* 22, 39–52.

Sheller, M. and J. Urry (2000) The City and The Car. *International Journal of Urban and Regional Reserch* 24 (4).

Sheller, M. and J. Urry (2006) The new mobilities paradigm. *Environmnet and Planning A* 38, 207–226.

Silva, E. (2011) Deliberate improvisation: Planning highway franchises in Santiago, Chile. *Planning Theory* 10, 35–52.

Susskind, L. and Elliott, M. (1983) *Paternalism, Conflict, and Coproduction: Learning from Citizen Action and Citizen Participation in Western Europe*, New York, NY: Plenum Press.

Thomson, K. (2001) *From Neighborhood to Nation the Democratic Foundations of Civil Society*. Hanover and London: University Press of New England.

United Nations Development Programme Annual Report (2007), *Making Globalization Work for All*, Office of Communications, United Nations Development Programme, New York.

Vasconcellos, E.A.D. (2001) *Urban Transport, Environment, and Equity: The Case for Developing Countries*. London, Sterling, VA: Earthscan Publications.

Venkatesan, V. (2010) *Licence to Live*. Available at: www.frontline.in/static/html/fl2705/stories/20100312270509900.htm.

2 Negotiating access

Urban planning policy and the social production of street vendor micro-mobilities in Hanoi, Vietnam

Noelani Eidse

Recent state-led efforts to develop and modernize Hanoi (the Socialist Republic of Vietnam's capital city) have sparked processes of rapid growth, initiating an urban transition that is reshaping the city's physical and socio-economic landscape. Hanoi's government is effectively redefining access to urban space in the capital through large-scale master plans and urban governance policies aimed at modernizing the city. In 2008, the municipal government introduced a street vending ban, prohibiting vendors from trading on 63 streets and in 48 public spaces. This policy privileges auto-mobility and positions vendors as obstructions to traffic flow and urban development. Considering the state's attempt to eliminate vending from the urban core, informal traders face increasingly precarious working conditions. Nonetheless, vendors do not receive this policy passively, and undertake careful negotiations to ply their wares and stake a claim to Hanoi's streets. Drawing on 18 months of ethnographic fieldwork, I assess the pressure on mobile vendors and explore the ways in which vendors push back against state sanctions. To understand the complexities of the contest for Hanoi's street spaces, I draw on a post-structuralist feminist approach, informed by scholarship on social spatialisations, mobilities and everyday politics. I explore vendors' lived realities and their connections with and contestations of local, regional and global political-economic systems. Set against the backdrop of a post-colonial, socialist cityscape undergoing rapid transformation, I find that mobility itself is a tool for resistance; by remaining on the move, vendors maintain their livelihoods and stake a claim to Hanoi's pavements, despite threats of exclusion.

Introduction

Within Hanoi, the capital of Vietnam (population 7.5 million), state-led efforts to modernize and polish the city have initiated rapid and unprecedented urban growth and transformation in recent years (Labbé, 2010). Through large-scale master plans and urban governance policies, Hanoi's government is effectively redefining access to urban space in the capital. As part of its plan to modernize the city, the municipal government introduced a ban on street vending in 2008, which prohibits vendors from plying their wares on 63[1] streets and from the vicinity of 48 public spaces (People's Committee of Hanoi, 2008). This policy forbids vendors from "blocking transportation" and reflects the state's approach to urban development. As is often the case in Global South contexts, increased auto-mobility is equated with progress; vehicle traffic is prioritized in state planning policy, while vendors plying their wares on street and sidewalk spaces are positioned as inhibitors to the flow of traffic, unproductive and obstacles in the "modern city" (Castells, 1996; Cross, 2000; Graham, 2014; Eidse et al., 2016).

In Hanoi this spatial discourse results in a contested urban landscape in which the everyday users of Hanoi's streets and sidewalks are at odds with official plans for urban development. Hanoi's urban governance strategies such as the 2008 vending ban hinge on exclusion, exerting disparate pressure on the micro-mobilities of residents, and threatening to displace and immobilize the city's informal vendors. Policies such as the 2008 vending ban disproportionately affect itinerant traders who often originate from the peri-urban fringe. In contrast to their local, stationary counterparts who draw on their social capital and perceived rights to the city's streetscapes, itinerant vendors are largely unable to claim a fixed selling location and are targeted more intensely by law enforcement officials as they carve out their living "on the move". Access to streets, sidewalks and public spaces is delineated in everyday life according to socio-economic positioning, gender and place of origin, among other factors. This poses significant challenges to those relying on street vending as an economic survival strategy. However, vendors are far from passive, and undertake careful negotiations daily to stake a claim to the city streets and pavements, avoid reprisals from the state and maintain their mobile livelihoods (Eidse and Turner, 2014; Eidse et al., 2016).

Building on previous ethnographies of vending livelihood practices in Hanoi (c.f. DiGregorio, 1994; Drummond, 1993; Eidse et al., 2016; Higgs, 2003; Jensen et al., 2013; Koh, 2008; Lincoln, 2008; Mitchell, 1995; Tana, 1996; Turner and Schoenberger, 2012), this chapter contributes an in-depth examination into the contradictions and coherences between state planning and vendors' everyday (im)mobilities and survival strategies. I assess the pressure exerted on mobile vendors and explore the ways in which vendors push back in order to maintain their livelihoods despite the threat of state sanctions. To understand the complex factors at play in the contest for Hanoi's street spaces, I draw on a post-structuralist feminist approach, informed by scholarship on social spatialisations, mobilities and everyday politics. In so doing, I explore these vendors' daily, lived realities as well as their connections with and contestations of local, regional and global political-economic systems. Set against the backdrop of a post-colonial, socialist cityscape undergoing rapid transformation, I find that mobility is a tool for resistance; by remaining on the move, vendors strive to maintain their livelihoods, and indeed stake a claim to Hanoi's pavements despite threats of state sanctions and exclusion.

This study builds on 18 months of fieldwork carried out between 2010 and 2016, drawing on a primarily qualitative mixed-methods approach. My findings are informed by participant observation in urban and peri-urban Hanoi; semi-structured interviews with 20 law enforcement officials and policymakers;

conversational interviews with 465 street vendors, including stationary and itinerant traders; semi-structured interviews with 20 long-time residents of Hanoi; solicited diaries with 10 vendors (Eidse and Turner, 2014); and photovoice with 10 vending participants. In the remainder of this chapter, I provide a discussion of the conceptual underpinnings for my study, contextualize street vending in Hanoi and explore the impacts of planning policy on the daily micro-mobilities of the vendors at the heart of this piece.

Constructing a capital: a recent history of planning and expansion in Hanoi

Following the introduction of Đổi Mới[2] (economic reforms initiated in 1986), Vietnam's major cities (Ho Chi Minh City and Hanoi) have seen an unprecedented expansion (Waibel, 2004). In addition to economic restructuring which signals Vietnam's move towards a market-oriented economy, the impact of Đổi Mới is visible in the urban built environment, as state-led and private development has taken a distinctly pro-urbanization approach (DiGregorio, 2011; Labbé, 2011). The government's focus on city expansion is especially evident in the case of Hanoi (Duan and Mamoru, 2009; Leaf, 2002)

In the 1990s, the Hanoi municipal government began to expand the city's reach by annexing nearby rural villages, absorbing them into the capital's territory (Labbé, 2011). More recently in 2008, the administrative boundaries of Hanoi were redrawn, resulting in a significant transformation of the capital city (DiGregorio, 2011). Overnight, Hanoi expanded to an area of 3345 square kilometres, from a previous 992 square kilometres, resulting in a population that was nearly doubled in size, growing from approximately 3.5 to 6.2 million (Prime Minister of Vietnam, 2008; "Experts surprised by audacity of proposed Hanoi master plan" 2009). The extension of Hanoi's administrative boundaries carried strategic importance for the government, and was described by the president of the Vietnam Urban Planning and Development Association as "a step in the process to turn the capital city into the nation's centre for politics, economics, culture and international cooperation" (Talk Vietnam, 2014). Although Hanoi is currently the second largest city in Vietnam, following Ho Chi Minh City, Hanoi's 2008 expansion is part of a larger campaign by city authorities to develop and modernize the city (Turner and Schoenberger, 2012; "Supersized Hanoi", 2008; PPJ, 2011).

The state's modernization discourse is represented most recently by the master plan called "Hanoi Capital Construction Master Plan to 2030 and Vision to 2050"[3], which was officially approved by Vietnam's prime minister in 2011 (PPJ, 2011). This plan outlines the push for modernity as state-led efforts aim to reimagine and rebrand the capital by rethinking land management, public space, urban management and transportation. In particular, through its master plans, the state promotes urban densification and infilling, peri-urban development, and the construction of new urban areas as well as the urbanization of rural areas (DiGregorio, 2011; MoC, 2014). In addition to large-scale imperatives aimed at modernizing the capital city, and positioning Hanoi as a major economic centre (both nationally and regionally), the master plan comprises numerous sub-plans regarding transportation, water systems, markets and other infrastructural components of the urban landscape (PPJ, 2011). While prioritizing population growth, with the goal of reaching 10 million by 2030, the plan simultaneously aims to transform Hanoi into the world's first sustainable city[4] by the same year (PPJ, 2011; Turner, 2009). In an attempt to maintain significant green space in the city's design, 60 per cent of the capital region is allocated for a green corridor, affecting rich agricultural regions such as Hà Tây, a former province that is now part of the capital. The remaining 40 per cent is allocated for urban development. These ongoing efforts hinge on the conversion of agricultural land to urban districts (cf. DiGregorio, 2011; Labbé, 2010; 2011; McGee, 2009).

Areas now located within the city boundaries, previously designated for agriculture, are being reallocated for development and industrialization, which has increased pressure and limitations on the livelihoods of peri-urban residents (Labbé, 2010). Peri-urban areas, which Friedmann (2011: 428) describes as "spaces of encounter, conflict and transformation in the course of becoming urban", are undergoing transformative economic, cultural and political processes, affecting stakeholders in different forms and to varying degrees, and giving rise to countless property disputes. (DiGregorio, 2011; Labbé and Boudreau, 2011). Hanoi's peri-urban areas have increasingly become "zones of exclusion that bring a loss of identity for some people and new patterns of consumption for others" (Labbé and Boudreau, 2011: 288). While the landscape becomes marked by high-rise development projects, and up-scale condo developments, and expanding transport links, peri-urban residents around the city are under threat of losing their land (DiGregorio, 2011). In this way, Hanoi's expansion projects in peripheral regions take on drastically different meanings depending on the interests being considered (Friedmann, 2011). The state posits that this transformation is a necessary step towards Hanoi's modernization, while many residents see it as a disruptive force; urban expansion threatens to erase the villages that once scattered the city's periphery and challenges the very ability of peri-urban villagers to make a living (Labbé, 2010; MoC, 2011).

As livelihoods come under pressure in the peri-urban fringe, many have turned to Hanoi's informal sector to get by (Jensen, Peppard and Thang, 2013). Yet, just as Hanoi's peri-urban areas are the sites for competing interests, public space in the capital city has become the site of collision between state goals for development and modernization and the needs of informal traders trying to carve out a living in Hanoi. In addition to the massive re-scaling of the city's territory and the resultant re-purposing and functional reorganization of Hanoi's peri-urban areas, state master planning initiatives outline several key changes that would disrupt internal neighbourhoods, such as the construction of a major highway coming in from east of the Red River, which threatens to raze numerous urban neighbourhoods (Turner and Schoenberger, 2012; Waibel, 2011). What results is an urban landscape thrust into a state of rapid transformation, with state plans for the capital city threatening to erase the lived spaces of its citizens.

Symbolic streetscapes: the prioritization of automobility over informality

In Hanoi, street spaces become symbols of progress and modernization in state development narrative. The contest for the city pavements is made evident by state attempts to reimagine, reorder and regulate street spaces as a means of increasing global connectedness (Eidse et al., 2016). The state's fixation with automobility made evident by the ban mirrors a longstanding rationale in Global South contexts that equates progress with fluid movement, prioritizing "spaces of flow" over "spaces of place" (Castells, 1996; Khayesi et al., 2010). State-initiated transport links – including highways, expressways and a metropolitan railway system – all privilege modern mobilities and gesture towards patterns of disparity emerging alongside increased urbanization (Smart and Smart, 2003; Cresswell, 2006). Policies that reimagine streets as channels for the efficient flow of traffic are introduced at the cost of the everyday social character of the street, neglecting the everyday users of street spaces such as pedestrians, cyclists and street vendors (Khayesi et al., 2010).

As is often the case in Global South contexts, urban governance and planning strategies position informal vendors as "out of place" on streets and sidewalks, and threaten their access to public space for their trade (Bromley, 2000; Brown, 2006; Eidse et al., 2016; Hansen et al., 2013). Phi Thai Binh, vice-chairman of Hanoi's People's Committee at the time of the ban's introduction, described the motivation behind the ban as an effort to "re-establish urban order in a civilized way" (Viet Nam News 2008). The modern city is imagined as a site for automobility, and as a result streets become representational spaces linked to the assertion of national identity (Short and Pinet-Peralta, 2010). Within Hanoi's contested urban landscape, numerous stakeholders compete for use of the streets. As Hanoi's urban planners design streets as spaces of flow, they threaten and displace the spaces of exchange, or "spaces of place" that are central to the social, economic and political liveability of the streetscape (Appleyard, 1981; Hutabarat, 2010; Khayesi et al., 2010). Set against the backdrop of the city streetscape, the competition between imagined and lived mobilities is underpinned by inequity. Everyday users who are categorized as being "out of place" – such as vendors – are neglected by vehicle-centred planning initiatives, and are presented as obstacles to the flow of traffic, and by extension to progress itself (Short and Pinet-Peralta, 2010).

Vendor experiences of (im)mobility in Hanoi

Street vending is one of the most visible means by which thousands of inhabitants make a living in Hanoi, and it is a livelihood that can have the least start-up costs as well (Gorman, 2008). The most recent data available notes that informal selling in Hanoi offers more than 11,000 people a livelihood means through marketplace or street vending (M4P, 2007). Before the economic renovations, which occurred under Đổi Mới in the mid-1980s, city pavements and streets were seldom used for street vending. However, with these economic reforms, Hanoi's streets became increasingly active, with countless vendors. These vendors can be either itinerant, or fixed selling from small stalls on pavements. The majority of vendors, especially those trading itinerantly, are rural to urban migrants, often women, who lack access to more durable livelihoods because of limited formal education, financial capital or social networks (Eidse et al., 2016; Higgs, 2003; Mitchell, 2008). They pursue this trade in response to changing conditions in their home villages, including decreasing land access and a growing inability to survive on farming alone, and are often the primary income earners for their households. Duong, a mobile vendor selling tourist goods explained: "I work this way in order to support my family – to take care of my husband, and to pay for our farming costs. It is important to us to keep farming, but it doesn't provide us with any income. Every day I sell in Hanoi to pay our bills and buy food for my family, and so I can give little things to my grandchildren". For vendors like Duong, who primarily cultivate rice for personal consumption, the bustling streets of urban Hanoi provide a much-needed site for economic survival.[5]

Regardless of the livelihoods vending provides, recent government efforts have sought to restrict the use of public space for such trade. As part of its plans to modernize and develop Hanoi, the municipal government has banned street vending on 62 selected streets and from 48 public spaces since 2008 (People's Committee of Hanoi, 2008). The contest for Hanoi's streets is undertaken daily, as the government imprints its modernization discourse on the urban landscape, exerting pressure on the lived spaces created by the city's informal traders. By trading on, or even near the officially banned sites, vendors run the risk of being arrested, fined or having their goods and equipment confiscated. Lien highlighted the tug-of-war between the municipal government and vendors like herself, who struggle to conduct their business: "It is the government's business to ban vending – to make the streets clean and organized – but it is my business to make a living so I can support my family". Lien's statement highlights the multiple interests involved in the competition for the city pavements.

As state planners design Hanoi's streets as spaces of flow, they threaten and displace the spaces of exchange, or "spaces of place" that are central to vending livelihoods. One urban planner working in Hanoi noted: "Street vendors who take up space in the streets or on the sidewalks disrupt other people from being able to use that space for transportation – to walk or drive their motorbikes. [Vending] is not the intended use, so it is not allowed". Many vendors have internalized these narratives. When asked why vending is banned, Siu, a vendor who uses a bicycle to sell seasonal fruit noted: "I think I am not allowed to sell because I take up too much space, I block the traffic . . . The cars might take up more space, but I stop more than they do. But what can I do? I do this because I don't have any other way to get by". In my case study, the enforcement of the 2008 vending ban, aiming to reduce "slow mobility", has mainly targeted itinerant traders, who overwhelmingly originate from Hanoi's peri-urban zones (cf. Agergaard and Thao, 2011; Jensen et al., 2013).

While the state's mobile imaginary draws distinct lines between allowable uses of street and sidewalk spaces and vending, mobility hierarchies are likewise evident between stationary and itinerant vendors. Access to streets, sidewalks and public space is delineated in everyday life according to socio-economic positioning, gender and place of origin, among other factors, which exerts significant pressure on those who rely on street vending as their primary strategy for economic survival. Itinerant vendors experience pressure to "keep moving" from local residents who exert their claim to the pavement. One mango vendor, Linh, explained that "we sometimes rest in front of that pharmacy during lunch, but if we don't move quickly enough when they open for business again, they will curse, push our carts into the street, throw tea in our faces and kick us". Vendors are compelled to stay "on the move" and pause cautiously; they are always ready to pick up their goods and carry on. Itinerant vendors remain on the move almost constantly while stationary, local vendors often have social connections with local authorities and draw on a perceived right to the city space. One young shoe vendor explained: "Locals have more rights to the sidewalk than we [migrant vendors] do – they wouldn't sell itinerantly, because they are able to set up stalls, and we wouldn't dare sell in one place". Vendors likewise experience pressure to keep moving, even when they are not selling.

Mobility is particularly important for itinerant vendors, given that their customer bases are relatively dispersed. By moving from place to place, vendors are able to utilize their mobility, reaching a greater number of clients throughout the day. Explaining the pressure, she feels to keep moving, an itinerant DVD vendor, Huynh, wrote in her journal: "I get so tired from walking around, but the longer I sit in one place, the more sales I miss". Huynh's statement also highlights the physical toll of remaining on the move. Similarly, Thanh, a young lighter vendor, explained that in addition to the difficulties he faces trying to make a steady income, he pays a physical price for his working itinerantly: "My tray of lighters is really heavy, and the strap sometimes leaves bruises on my shoulder and neck. I often get sunburnt from working outside, and by the end of the day my feet are sore from the hours I spend walking around". Thanh's statement highlights the everyday challenges faced by itinerant vendors, who lack the social and financial capital needed to secure fixed trading sites, and must remain on the move in order to carve out a living. Itinerant vendors must navigate power imbalances between themselves and the state, while likewise competing with residents and fixed vendors for access to the pavements. Despite intense competiton for Hanoi's pavements—compounded with the potential consequences of trading on the edge of legality—. vendors are not deterred from working in Hanoi, and can still be seen plying their trade throughout the city.

Conceptualizing the contest for Hanoi's pavements: social spatializations and the everyday politics of vending (im)mobilities

> "Space is not just the place of conflict, but an object of the struggle itself."
> (Elden, 2004: 183)

Although scholars have used a range of frameworks to conceptualize competition for, and claims to space over the years, the notion of social spatializations introduced by Lefebvre has laid the groundwork for exploring the nuanced social dimensions of space (Lefebvre, 1991; De Certeau, 1984; Shields, 1991; Soja, 1989). Lefebvre (1991) understands space as a social product, composed of three parts: that which is imagined, lived and perceived.[6] This spatial triad offers a platform for deconstructing the ways in which space is socially produced and contested. The politics of spatial access are made particularly evident by the interplay of imagined and lived space. Imagined space is designed by those in positions of power – such as architects, engineers, urban planners or government representatives – and refers to the conceptions of how space *should* be. It hinges on binaries and oppositions, delineating between proper and improper uses of space (Jacobs, 1993). Often drawn according to *a priori* categories, imagined space frequently misses the nuances of what is actually occurring on the ground (Blomley, 2011). In other words, those occupying positions of authority who imagine or plan the urban landscape often purposefully disregard the ways in which the greater population uses that very space – excluding so-called anti-modern everyday users of urban space in order to attract more desirable users matching the states' modernization narrative (McCann, 1999; Merrifield, 2006).

As the protagonists of imagined space attempt to mould and engineer an area, efficiency is chosen over the chaos of everyday life, and pre-existing lived space is threatened with destruction (Lefebvre, 1991; Jacobs, 1993). Although street spaces function as an integral livelihood site for those on the economic fringe, such as informal vendors, official approaches to urban governance often aim to repurpose these spaces for fluid movement. While Hanoi's municipal government demonstrates a distinctly exclusionary stance on itinerant traders, other urban locales in Southeast Asia – such as Bangkok, Kuala Lumpur and Manila – have taken an approach of planning for the coexistence of flow and informal trade; in so doing, these cities recognize street vending as an important element of the urban fabric, suggesting that a city can foster both fluid mobility and spaces for trade (Bhowmik, 2006; Yasmeen and Nirathron, 2014). Still, in Hanoi, the interests of the state and residents clash as locales such as the city pavements – "lived spaces" – which serve significant functions for the general public, come under the authority of the government-represented imagined space (Blomley, 2011; Leach, 1976). What is more, those who produce lived space have comparatively limited economic and political power, and therefore are challenged in their ability to maintain control of their space (Lefebvre, 1991).

Through an exertion of "symbolic materiality", by which spaces are equated with abstract ideas, elements of the urban landscape are transformed into ideological symbols (Shields, 1991). As imagined space is used to exert state modernization narratives, the space itself becomes a tool for domination, revealing complex power dynamics at play (Jacobs, 1993; Leach, 1976; Merrifield, 1993, 2006). Indeed, in Global South cities, state attempts to control street spaces often become entangled with development discourses (van Blerk, 2013). State urban planners imagine clear and organized streets that facilitate the flow of traffic, equating improved automobility as a signal of modernity (Khayesi et al., 2010). On the other hand, streets that are bustling with informal trade and transport services become harbingers of the past and are equated with the traditional, and as a result they are often viewed as a hindrance to development (Koh, 2004; Lincoln, 2008). As a result of these sentiments, the use of streets and sidewalks for informal trade is presented as "private encroachment of public space" (Blomley, 2011: 13). Space takes on a distinctly political meaning, as both the everyday users, and those who occupy authoritative positions, attempt to exert their demands through, and upon the space itself (Elden, 2004; Habermas, 1989; Howell, 1993). Through the use of symbolic materiality, the vision and ideals of the few in positions of power are imposed on the lived realities of the vast majority of the population (Blomley, 2011).

State-led efforts to "imagine" and shape the urban landscape itself are entangled with attempts to shape everyday mobilities. The tension between lived and imagined space highlights the underlying politics at play in the struggle for space, and mirrors the dissidence between the everyday micro-mobilities of street vendors and the state's attempt to cultivate "modern" mobilities. As those in positions of power attempt to shape and engineer space to fit an ideal, the ways in which bodies move through space comes under pressure as well. The concept of mobility facilitates critical discussion of the politics and power dynamics at play in processes of movement, raising questions about who is and is not able to move, what forms of movement are privileged and desired over others, and how the same movement can take on drastically different meanings depending on the positionality of the mobile subject and the motive force behind their movement (Adey et al., 2014; Cresswell, 2006, 2010; Sheller and Urry, 2006; Tanzarn, 2008; Urry, 2000; Uteng, 2009). As such, the contributions of mobility scholarship centre on its deconstruction of the meaning of movement (Blomley, 2014; Graham, 2014; Jensen, 2009; Oswin, 2014). The concept of mobility provides a platform for addressing the ways in which movement is socially produced and reworked, inherently political and differential (Calvo, 1994; Tanzarn, 2012). Both produced by, and reproductive of, hierarchies of power and social exclusion, for some privileged few mobility comes at the cost of the mobility of the majority of "others"; referred to as mobility injustice, the production of mobile hierarchies reinforces the gap between (im)mobile subjects (Cresswell, 2010; Sheller, 2008). When delineating between acceptable and unacceptable mobilities, those in positions of power demonstrate a rationale of functionality, privileging modern, fluid mobilities over informal, traditional mobilities (Cresswell, 2006). In the case of state interventions and urban planning in Hanoi, so-called "modern" forms of transportation, such as private motorized vehicles, are privileged and encouraged in the name of progress, while "traditional" forms of mobility, such as roving street vendors – are viewed as stagnant and anti-modern, and are thus discouraged (Blomley, 2014; Graham, 2014). Those who do not conform to "acceptable mobilities" become obstructions to the flow, in the eye of the state, rather than part of the flow. As a result, the capital city's itinerant vendors are denied the right to be immobile – even though at times remaining stationary could be beneficial for their business.

Hanoi's urban planning trajectory is highly exclusive, drawing divisive lines between those who are and are not "allowed" to stake a claim to the city's future. Street vending is presented by the state as "anti-modern", uncivilized and a hindrance to development – a characterization used to justify restrictions on the everyday mobilities of Hanoi's informal traders. Leshkowich (2014) argues that attempts to render vending livelihoods as chaotic, disorderly and problematic are highly gendered and closely linked to attempts at regulating and controlling female mobility. In other words, street vendors are targeted by state reprisals because of who they are, rather than how they make a living. This is clear in the case of street vending in Vietnam, an activity predominantly undertaken by female labourers,[7] who lack access to more durable livelihood options because of a lack of formal education, financial and social capital; women make up the majority of vendors, and are thus affected most directly by regulations on informal urban trading (Eidse et al., 2016). To understand the underlying factors contributing to the everyday politics and disparate outcomes of Hanoi's urban development, one must therefore consider positionality, addressing how the urban mobilities of informal traders are embodied and gendered.

McCann (2011: 121) argues that "mobility is stratified and conditioned by access to resources and by one's identity". Scholars have increasingly considered this interplay between identity and mobility, deconstructing the ways in which mobilities are tethered to positionality (including gender, ethnicity, socio-economic standing and occupation among other elements) (Graham, 2001; Srinivasan, 2008; Tanzarn, 2012; Turner and Oswin, 2015; Ureta, 2008; Uteng, 2009). These categories are far from static, but rather are continually produced, reworked and reconfigured through social relations (Butler, 1988; Ellis, 2000; Haraway, 1991; Reddock, 2000). Identity categories provide a means for examining the ways in which power dynamics and positionality frame the individuals' access to assets and activities, and in turn their everyday mobilities (Arun, 2012; Reddock, 2000; Bonnin and Turner, 2014; Tanzarn, 2012).

The power imbalance between mobile subjects can be understood according to Cresswell's (2010) "politics of mobility", which highlights how experiences of mobility take on drastically different meanings from subject to subject. Situated within a discussion of positionality, the concept of mobility provides a means for interrogating disparate power dynamics and differential access (Hanson, 2010; Tanzarn, 2012). Development and livelihood studies likewise emphasize the importance of integrating gender awareness into explorations of livelihood practices, providing a platform to critically assess the ways in which livelihood activities are engendered (Barriteau, 2000; Boserup, 1970; Ellis, 1993, 1998; Little, 2004; Radel, 2012; Scott, 1986). While identity and positionality significantly impact individuals' access to livelihood opportunities and resources, few studies have addressed the interplay between gender, mobility and livelihoods in the Global South (Agarwal, 1997; Bonnin and Turner, 2014; Carney, 1998; Dolan, 2002). In using a gender lens to better understand itinerant vending mobilities, I am able to tease apart the gendered asymmetries regarding state decision-making and planning, access to public space, and informal sector livelihoods and contextualize local gender dynamics within the political economy in the Global South (Dalla Costa and Dalla Costa, 1995; Hanson, 2010; Haraway, 1991; Agarwal, 1992).

Staking a claim to Hanoi's streets: the everyday politics of vendor (im)mobilities

Through a discussion of vendors' everyday politics, I am able to better understand the daily actions vendors take in order to push back against state structures that threaten to immobilize them. Kerkvliet (2009: 232) describes everyday politics as "people embracing, complying with, adjusting, and contesting norms and rules regarding authority over, production of, or allocation of resources and doing so in quiet, mundane, and subtle expressions and acts that are rarely organized or direct". In contrast to conventional political studies, which tend to focus on official politics of organizations and authorities, everyday politics highlight the political underpinnings of daily actions (Havel and Wilson, 1985; Kerkvliet, 2009). For itinerant vendors trading on the edge of legality, daily micro-mobilities become a tool to contest state-led planning in Hanoi; in other words, their mobility itself becomes a means of resistance (Eidse et al., 2016; Scott, 2009).

In response to state-initiated restrictions on their trade, vendors must undertake careful negotiations daily to maintain their trade. Considering the state's attempt to eliminate street vending from the urban core, vendors are faced with increasingly precarious working conditions. Restrictions such as the 2008 ban compound with an already tenuous working environment, requiring vendors to undertake covert resistance tactics in order to ply their wares. Harnessing the benefits of their itinerancy, vendors strategically adjust their daily rhythms to avoid reprisals for their trade. An itinerant vegetable vendor using a carrying pole explained:

> I keep moving during the mornings and only stop when I'm making a sale. But during lunchtime, the police usually rest for a couple of hours, so I can rest then too. I try to find shady spots in front of closed shops to sit where I won't bother anyone. Then I pick up and keep moving again.

Vendors are savvy to the daily and weekly rhythms of ward officials and police, and have created micro-mobility patterns to avoid fines and retributions. Police repeatedly patrol at fixed times during the day, and are not particularly inventive in their routines. Vendors build on these repeated customs to create their own trade mobilities. By resting when the police are less active and moving when the police patrol, they minimize the physical strain of their mobile livelihoods while also reducing their interaction with police. Staying "on the move" is central to vendors' everyday resistance tactics, since it is the very fact that they are itinerant that helps them avoid the consequences of trading.

Itinerant vendors experience restricted mobilities in the city, both when they are actively trading or simply walking through the city streets. Siu, a mobile vendor selling tourist goods noted: "When the police are out, even if I am just walking

I will get a fine – so I don't go out then". Because the police where Siu trades are familiar with her, she is easily recognizable as a vendor, despite her discrete set-up, consisting of a small purse, which she uses to carry bracelets and baseball caps. For itinerant vendors, movement is restricted because of who they are, not what they are doing (Leshkowich, 2014). Simply having sellable wares in their possession can be grounds for a fine. To navigate these restrictions on their mobilities, vendors cultivate relationships with other vendors, as well as sympathetic residents or other individuals undertaking different forms of work nearby. By making connections in the neighbourhoods where they sell, vendors are able to store their goods for short periods during police raids. One noodle vendor, Ly, explained that she constantly keeps an eye out for approaching police, and quickly moves her carrying pole into a resident's home during raids: "There is a lady that lives close to one of my regular selling spots. Whenever the police are doing a sweep of the street [clearing vendors], she lets me move in to hide". By cultivating social capital with residents, vendors are able to negotiate the limitations on their trade.

Still other vendors manage to harness the benefits of their itinerancy, using their very mobility as a tool for resistance. In her journal, one young balloon vendor, Nam, described how she and her friends sell on a bustling banned street, but make use of the restrictions on police mobilities to flee during raids: "We go into small streets, small alleys where they cannot go in with their trucks, or we run to one-way streets where trucks cannot come in either". Additionally, vendors adjust their routes in order to resist the ban. Explaining his decision to take a short-cut during police crackdowns, Thanh, the lighter vendor introduced earlier, noted:

> I take this route because it's hidden, I can stay out of sight from the police when they are having a campaign to clear the streets [enforcing the ban strictly]. Normally I would prefer to walk the long way and meet my regular customers, but during campaigns, even if vendors are not selling, even if they are just walking home, they can still get arrested, so that's why this shortcut helps me to be less vulnerable to the police.

In response to the state's attempts to eliminate vending in Hanoi, actively excluding them from plans for a new Hanoi, the ways in which vendors use public space takes on a distinctly political meaning. By adjusting their daily rhythms and routines, vendors undertake everyday resistance in order to carve out a place in Hanoi's development. By adjusting their rhythms and routines vendors continually renegotiate their access to the city's streets, manage to carve out a living, and stake claim to Hanoi's future.

Policy implications and concluding remarks

As Asia catches the global spotlight in relation to its economic growth while other regions of the globe economically contract, the expanding disparities in livelihood possibilities in Vietnam have become more and more pronounced(Sheppard, 2012). One of the results is an urban crisis in which burgeoning economic expansion is paired with rising inequality and pressure on local livelihoods. These trends are clear within the rapidly transitioning urban areas of Hanoi, where Vietnam's socialist state is struggling to mix development goals aimed at market integration with a continued tight control over society. Hanoi's development can be understood as a process founded on a distinct binary impressed upon the urban landscape; while the city's economic growth is praised and indeed fostered by the state, on the ground, Hanoi's accelerating expansion has thrust thousands of residents into a state of crisis (DiGregorio, 2011). The dilemma is that as the municipal government introduces policies that favour industrialization, development and the formal sector, thousands of residents are actively excluded from the city's future, and their livelihoods are put on the line. Hanoi's municipal government seeks to increase automobility, while curtailing so-called "anti-modern" everyday mobilities, threatening to displace the thousands of vendors who ply their trade in Hanoi's street and sidewalk spaces.

Amidst growing inequity associated with urban development, and the cost of living increasing despite static agricultural profits, the informal sector provides an important opportunity for entrepreneurship and economic survival (Hansen et al., 2013; Rigg, 2004; Sassen, 1994). Informal labourers, such as street vendors, respond to market demands in a way that secures their own livelihoods, while offering convenient services to consumers (Turner, 2009). Yet still, in Hanoi state initiatives threaten to eliminate the informal sector while failing to address the complex factors contributing to its continued presence (Jensen and Peppard, 2003).

Within this chapter, I have illustrated the complex range of impacts that Hanoi's municipal planning has exerted on the daily micro-mobilities of its residents, all in the name of "progress" and "development". Furthermore, I have explored the ways in which vendors – who are actively excluded and targeted by state planning policy – have continued to carve out their livelihoods on Hanoi's sidewalks and pavements. In Hanoi, state development and modernization initiatives are continually reshaping mobilities. Those occupying positions of power demonstrate a rationale of functionality when demarcating boundaries between acceptable and unacceptable mobilities, privileging modern, fluid mobilities above those that are traditional and staccato. As vendors carve out informal everyday mobilities, they are seen by the state as obstructions to, rather than part of the flow, as demonstrated by the 2008 vending ban. Nonetheless, vendors do not receive this policy's implementation passively, and find ingenious ways to circumvent restrictions on their trade.

Informal mobilities are intricately connected to processes of negotiation and contestation, as vendors carry out forms of movement beyond the purview of the state. In response to accelerated urban development and diminishing livelihood options in Hanoi's peri-urban fringe, itinerant migrant labourers are turning to Hanoi's informal sector – notably, street vending – in order to get by, despite the state's attempt to restrict street vending livelihoods. Moreover, street vendors are able to continually carve out a trading space because of the very fact that their mobilities are informal and exist beyond state regulation. Power dynamics are at play here that underpin the hierarchies of informal trade in the Global South, and street vendors, members of the "kinetic under-class", have found ways to push back against the very structures that seek to immobilize them, striving to maintain mobile livelihoods despite threats of state sanctions and exclusion.

Notes

1 Although the 2008 vending ban prohibited vending on 62 streets, the policy was revised in 2009 to include an additional street, resulting in a total 63 banned streets. The updated ban can be viewed online at http://m.thu vienphapluat.vn/archive/detail/84303.
2 Đổi Mới can be understood as follows: "At its simplest, Đổi Mới involves a transition from a centrally planned to a market-based, 'multisectoral' economy, in which household enterprises, private businesses, foreign firms and joint ventures are allowed to operate as autonomous entities alongside state-owned enterprises and cooperatives" (UNDP 1998: 1).
3 Master plans are prepared by the Vietnam Institute of Architecture and Planning every five years.
4 Hanoi's 2030 plan draws on four "pillars of sustainability": economic, environment, cultural and social (Viet Nam Net 2009).
5 Within Vietnam there is a decreasing profit margin for rice cultivation, with smaller land parcels available for farming and increasing cost of inputs while market prices for rice remain much the same. Still, vendors expressed their preference to eat rice that they grow themselves, since they have greater control over the varieties grown and inputs used, and further explained that farming is a core element of their identities which they aim to preserve. However, although residents of peri-urban and rural areas continue to engage in farming, they must seek income-generating activities elsewhere – most often in the city – to meet their daily living costs.
6 While perceived space corresponds with the features, structures and material constraints that form physical space, lived space is created through everyday use, and imagined space reflects the ways in which space is designed and planned (Merrifield 1993).
7 Because of the excess labour force in rural areas and the lack of adequate land for agricultural production, women frequently leave their rural households to pursue income-generating activities in the city, sending remittances back to their rural households, whereas men often remain in rural areas to manage farming responsibilities and care for their children (Tiến and Ngọc 2001).

References

Adey, P., Bissel, D., Hannam, K., Merriman, P. and Sheller, M. (2014) Introduction. In: Adey, P., Bissel, D., Hannam, K., Merriman, P. and Sheller, M. (eds.), *The Routledge Handbook of Mobilities*, 1–20. Abingdon: Routledge.

Agarwal, B. (1992). The gender and environment debate: Lessons from India. *Feminist Studies* 18 (1), 119–158.

Agarwal, B. (1997). 'Bargaining' and gender relations: Within and beyond the household. *Feminist economics* 3 (1), 1–51.

Agergaard, J. and Thao, V.T. (2011) Mobile, flexible, and adaptable: Female migrants in Hanoi's informal sector. *Population, Space and Place* 17, 407–420.

Appleyard, D. (1981) Place and non place. In: de Neufville, J. I. (ed.), *The Land Use Policy Debate in the United States*, 49–55. New York: Springer.

Arun, S. (2012) 'We are farmers too': Agrarian change and gendered livelihoods in Kerala, South India. *Journal of Gender Studies* 21 (3), 271–284.

Barriteau, V.E. (2000) Feminist theory and development: Implications for policy, research and action. In: Parpart, J.L., Connelly, M.P. and Barriteau, V.E. (eds.), *Theoretical Perspectives on Gender and Development*, 161–178. Ottawa: International Development Research Centre.

Bhowmik, S. (2006) *Hawkers in the Urban Informal Sector: A Study of Street Vendors in Six Indian Cities*. India: National Alliance of Street Vendors of India.

Blomley, N.K. (2011) *Rights of Passage: Sidewalks and the Regulation of Public Flow*. Milton Park, Abingdon, Oxon; New York, NY: Routledge.

Blomley, NK. (2014) Sidewalks. In: Adey, P. (ed.), *The Routledge Handbook of Mobilities*. Oxon: Routledge.

Bonnin, C. and Turner, S. (2014). 'A good wife stays home': Gendered negotiations over state agricultural programmes, upland Vietnam. *Gender, Place and Culture* 21 (10), 1302–1320.

Boserup, E. (1970) *Women's Role in Economic Development*. London, UK: Allen and Unwin.

Bromley, R. (2000) Street vending and public policy: A global review. *International Journal of Sociology and Social Policy* 20 (1–2), 1–28.

Brown, A. (2006) Challenging street livelihoods. In: Brown, A. (ed.), *Contested Space: Street Trading, Public Space, and Livelihoods in Developing Cities*, 3–16. Warwickshire: ITDG Publishing

Butler, J. (1988) Performative acts and gender constitution: An essay in phenomenology and feminist theory. *Theatre Journal* 40 (1), 519–531.

Calvo, C. (1994) *Case Study on the Role of Women in Rural Transport: Access of Women to Domestic Facilities*. SSATP Working Paper No. 11. Washington, DC: World Bank.

Carney, D. (1998) *Sustainable Rural Livelihoods: What Contribution Can We Make?* London: Department for International Development.

Castells, M. (1996) *The Rise of the Network Society*. Oxford: Blackwell.

Cresswell, T. (2006) *On the Move: Mobility in the Modern Western World*. New York, NY: Routledge.

Cresswell, T. (2010). Towards a politics of mobility. *Environment and Planning D: Society and Space* 28 (1), 17–31.

Cross, J.C. (2000) Street vendors, modernity and postmodernity: Conflict and compromise in the global economy. *International Journal of Sociology and Social Policy* 1 (2), 30–52.

Dalla Costa, M. and Dalla Costa, G.F. (1995) Paying the price. *Women and the Politics of International Economic Strategy*. London: Zed Books.

De Certeau, M. (1984) *The Practice of Everyday Life*. Los Angeles: University of California Press.

DiGregorio, M. (1994) *Urban Harvest: Recycling as a Peasant Industry in Northern Vietnam*. East-West Center Occasional Papers No. 17. Hawaii: East-West Center.

DiGregorio, M. (2011). Into the land rush: Facing the urban transition in Hanoi's Western suburbs. *International Development Planning Review* 33 (3), 293–319.

Dolan, C. (2002) *Gender and diverse livelihoods in Uganda*. LADDER Working Paper No. 10. UK: Department for International Development; Norwich: University of East Anglia.

Drummond, L. (1993) *Women, the Household Economy, and the Informal Sector in Hanoi*. Unpublished Master's Thesis. Canada: University of British Columbia.

Duan, H.D. and Mamoru, S. (2009) Studies on Hanoi urban transition in the late 20th century based on GIS/RS. *Southeast Asian Studies* 46 (4), 532–546.
Eidse, N. and Turner, S. (2014) Doing resistance their own way: Counter-narratives of street vending in Hanoi, Vietnam through solicited journaling. *Area* 46, 242–248.
Eidse, N., Turner, S. and Oswin, N. (2016) Contesting street spaces in a socialist city: Itinerant vending-scapes and the everyday politics of mobility in Hanoi, Vietnam. *Annals of the American Association of Geographers* 106 (2), 340–349.
Elden, S. (2004) *Understanding Henri Lefebvre*. London: Bloomsbury Publishing,
Ellis, F. (1993) *Farm Households and Agrarian Development* (2nd ed.). Cambridge: Cambridge University Press.
Ellis, F. (1998) Household strategies and rural livelihood diversification. *The Journal of Development Studies* 35, 1–38.
Ellis, F. (2000) *Rural Livelihoods and Diversity in Developing Countries*. Oxford; New York, NY: Oxford University Press.
Friedmann, J. (2011) Becoming urban: Periurban dynamics in Vietnam and China. *Pacific Affairs* 84 (3), 425–434.
Gorman, T. (2008) Hanoi: Ban on street vending goes into effect July 1. *Open Air Market Network*. Available at: www.openair.org/node/431 (accessed 15 July 2010).
Graham, A. (2001) *Gender mainstreaming guidelines for disaster management programmes: A principled Socio-Economic and Gender Analysis (SEAGA) approach*. Paper presented in the expert group meeting on "Environmental Management and the Mitigation of Natural Disasters: A Gender Perspective", Ankara, Turkey, 6–9 November.
Graham, S. (2014) Disruptions. In: Adey, P., Bissel, D., Hannam, K., Merriman, P. and Sheller, M. (eds.), *The Routledge Handbook of Mobilities*, 468–471. Abingdon: Routledge.
Habermas, J. and Habermas, J. (1989) *The Theory of Communicative Action* (Vol. 2). Boston: Beacon Press.
Hansen, K.T., Little, W.E. and Milgram, B.L. (eds.) (2013) *Street Economies in the Urban Global South*. Santa Fe: School for Advanced Research Press.
Hanson, S. (2010) Gender and mobility: New approaches for informing sustainability. *Gender, Place and Culture: A Journal of Feminist Geography* 17 (1), 5–23.
Haraway, D. (1991) Situated knowledges. In: Haraway, D. (ed.), *Simians, Cyborgs, and Women: The Reinvention of Nature*, 183–201. New York, NY: Routledge
Havel, V. and Wilson, P. (1985) The power of the powerless. *International Journal of Politics* 15 (3/4), 23–96.
Higgs, P. (2003) Footpath traders in a Hanoi neighbourhood. In: Drummond, L. and Thomas, M. (eds.), *Consuming Urban Culture in Contemporary Vietnam*, 75–88. London and New York, NY: Routledge Curzon.
Howell, P. (1993) Public space and the public sphere: Political theory and the historical geography of modernity. *Environment and Planning D* 11, 303–303.
Hutabarat Lo, R. (2010) The city as a mirror: Transport, land use and social change in Jakarta. *Urban Studies* 1 (27), 1–27.
Jacobs, J.M. (1993) The city unbound: Qualitative approaches to the city. *Urban Studies* 30, 827–848.
Jensen, O. B. (2009) Flows of meaning, cultures of movements: urban mobility as meaningful everyday life practice. *Mobilities* 4 (1): 139–158.
Jensen, R. and Peppard, D. (2003) Hanoi's informal sector and the Vietnamese economy: A case study of roving street vendors. *Journal of Asian and African Studies* 38 (1), 71–84.

Jensen, R., Peppard, D. and Thang, V.T.M. (2013) *Women on the Move: Hanoi's Migrant Roving Street Vendors*. Hanoi: Women's Publishing House.

Kerkvliet, B.J.T. (2009) Everyday politics in peasant societies (and ours). *Journal of Peasant Studies* 36 (1), 227–243.

Khayesi, M., Monheim, H. and Nebe, J.M. (2010). Negotiating streets for all in urban transport planning: The case for pedestrians, cyclists and street vendors in Nairobi, Kenya. *Antipode* 42 (1), 103–126.

Koh, D. (2004) Urban government: Ward-level administration in Hanoi. In: Kerkvliet, B. and Marr, D. (eds.), *Beyond Hanoi: Local Government in Vietnam*, 197–228. Singapore: Nias Press.

Koh, D. (2008) The pavement as civic space: History and dynamics in the city of Hanoi. In: Douglass, M., Ho, K.C. and Ooi, G.L. (eds.), *Globalization, the City and Civil Society in Pacific Asia: The Social Production of Civic Spaces*, 145–174. London: Routledge.

Labbé, D. (2010) *Facing the Urban Transition in Hanoi: Recent Urban Planning Issues and Initiatives*. Institut National de la Recherche Scientifique Centre.

Labbé, D. (2011). Urban destruction and land disputes in peri-urban Hanoi during the late-socialist period. *Pacific Affairs* 84 (3), 435–457.

Labbé, D. and Boudreau, J.A. (2011) Understanding the causes of urban fragmentation in Hanoi: The case of new urban areas. *International Development Planning Review* 33, 273–291.

Leach, E. (1976) *Culture and Communication: The Logic by Which Symbols Are Connected*. New York, NY: Cambridge University Press.

Leaf, M. (2002) A tale of two villages: Globalization and peri-urban change in China and Vietnam. *Cities* 19, 23–31.

Leaf, M. (2011). Periurban Asia: A commentary on 'becoming urban'. *Pacific Affairs* 84 (3), 525–534.

Lefebvre, H. (1991) *The Production of Space*. Oxford: Basil Blackwell.

Leshkowich, A.M. (2014) *Essential Trade: Vietnamese Women in a Changing Marketplace*. Honolulu: University of Hawaii Press.

Lincoln, M. (2008) Report from the field: Street vendors and the informal sector in Hanoi. *Dialect Anthropol* 32, 261–265.

Little, W. E. (2004) *Mayas in the Market Place: Tourism, Globalization and Cultural Identity*. Austin: University of Texas Press.

M4P [Making Markets Work Better for the Poor] (2007) Street vending in Hanoi: Reconciling contradictory concerns. *Markets and Development Bulletin 13*. Available at: www.markets4poor.org/?name=publication&op=viewDetailNews&id=574 (accessed 6 July 2010).

McCann, E. (1999) Race, protest, and public space: Contextualizing Lefebvre in the US city. *Antipode* 31, 163–184.

McCann, E. (2011) Urban policy mobilities and global circuits of knowledge: Toward a research agenda. *Annals of the Association of American Geographers* 101 (1), 107–130.

McGee, T.G. (2009) Interrogating the production of urban space in China and Vietnam under market socialism. *Asia Pacific Viewpoint* 50 (2), 228–246.

Merrifield, A. (1993) Place and space: A Lefebvrian reconciliation. *Transactions of the Institute of British Geographers* 18, 516–531.

Merrifield, A. (2006) *Henri Lefebvre: A Critical Introduction*. New York, NY: Routledge.

Mitchell, C. (2008) Altered landscapes, altered livelihoods: The shifting experience of informal waste collecting during Hanoi's urban transition. *Geoforum* 39, 2019–2029.

Mitchell, D. (1995) The end of public space? People's park, definitions of the public, and democracy. *Annals of the American Geographer* 85 (1), 108–133.

MoC [Ministry of Construction] (2011) *Hanoi's Industries See Optimistic Growth*, 25 April. Hanoi, Construction Publishing House. Construction Information Center. Available at: www.moc.gov.vn (accessed 8 March 2016).

MoC [Ministry of Construction] (2014) *PM approves adjustments to Hanoi Capital Region Master Plan*. May 12. Hanoi, Construction Publishing House. Construction Information Center. Available at: www.moc.gov.vn (accessed 10 June 2016).

Oswin, N. (2014) Queer theory. In: Adey, P., Bissel, D., Hannam, K., Merriman, P. and Sheller, M. (eds.) *The Routledge Handbook of Mobilities*, 85–93. Abingdon: Routledge.

People's Committee of Hanoi (2008) 02/2008/QD-UBND Quyết định. Ban hành Quy định về quản lý hoạt động bán hàng rong trên địa bàn Thành phố Hà Nội [Decision. Promulgating regulation on management of street-selling activities in Hanoi]. Revised policy: No. 46/2009/QD-UBND 15 January 2009.

PPJ [Perkins Eastman-Posco E&C and JINA] (2011) *HUPI: Hanoi Master Plan to 2030 and Vision to 2050*. Hanoi: Ha Noi's Department of Planning and Architecture.

Prime Minister of Vietnam (2008) Quyết định 490/QĐ-TTg phê duyệt quy hoạch xây dựng vùng Thủ đô Hà Nội [Decision of the Prime Minister on the approval of construction planning for Hanoi capital zone].

Prime Minister of Vietnam (2008) Quyết định 490/QĐ-TTg phê duyệt quy hoạch xây dựng vùng Thủ đô Hà Nội [Decision of the Prime Minister on the approval of construction planning for Hanoi capital zone].

Radel, C. (2012) Gendered livelihoods and the politics of socio-environmental identity: Women's participation in conservation projects in Calakmul, Mexico. *Gender, Place and Culture: A Journal of Feminist Geography* 19 (1), 61–82.

Reddock, R. (2000) Why gender? Why development? In: Parpart, J.L., Connelly, M.P. and Barriteau, V.E. (eds.), *Theoretical Perspectives on Gender and Development*, 23–50. Ottawa: International Development Research Centre (IDRC).

Rigg, J. (2004) *Southeast Asia: The Human Landscape of Modernization and Development*. London: Routledge.

Sassen, S. (1994) The informal economy: Between new developments and old regulations. *Yale Law Journal*, 103 (8), 2289–2304.

Scott, J.W. (1986) Gender: A useful category of historical analysis. *The American historical review*, 91 (5), 1053–1075.

Scott, J. (2009) *The art of not being governed: An anarchist history of upland Southeast Asia*. New Haven; London: Yale University Press, 2009.

Sheller, M. (2008) Mobility, freedom and public space. In: Bergmann, S. and Sager, T. (eds.) *The Ethics of Mobilities: Rethinking Place, Exclusion, Freedom, and Environment*, 25–38. London and New York: Routledge.

Sheller, M. and Urry, J. (2006) The new mobilities paradigm. *Environment and Planning* 38 (2), 207–226.

Sheppard, E. (2012). Trade, globalization and uneven development: Entanglements of geographical political economy. *Progress in Human Geography* 36 (1), 4471.

Shields, R. (1991) *Places on the Margin: Alternative Geographies of Modernity*. London: Routledge.

Short, J.R. and Pinet-Peralta, L.M. (2010). No accident: Traffic and pedestrians in the modern city. *Mobilities* 5 (1), 41–59.

Smart, A. and Smart, J. (2003) Urbanization and the global perspective. *Annual Review of Anthropology* 32, 263–285.

Soja, E. (1989) *Postmodern Geographies, the Reassertion of Space in Critical Social Theory*. London: Verso.

Srinivasan, S. (2008) An exploration of the accessibility of low-income women: Chengdu, China and Chennai, India. In: Uteng, T.P. and Cresswell, T. (eds.), *Gendered Mobilities*, 143–158. Farnham: Ashgate.

The Straits Times (2008) Supersized Hanoi, 7 June. Available at: www.streetnet.org.za/index.html (accessed 5 June 2009).

Talk Vietnam (2014) Hanoi sets sights on becoming modern metropolis by 2030, 26 June. Available at: www.talkvietnam.com/2014/06/hanoi-sets-sights-on-becoming-modern-metropolis-by-2030/ (accessed 3 June 2016).

Tana, L. (1996) *Rural-Urban Migration in the Hanoi Region*. Singapore: Institute of Southeast Asian Studies.

Tanzarn, N. (2008) Gendered mobilities in developing countries: the case of (urban) Uganda. In: Uteng, T. P. and Cresswell, T. (eds.) *Gendered Mobilities*. Aldershot: Ashgate, 159–172.

Tanzarn, N. (2012) Gendered mobilities in developing countries: The case of (urban) Uganda. In: Uteng, T.P. and Cresswell, T. (eds.), *Gendered Mobilities*, 159–172. Burlington: Ashgate Publishing.

Tien, H.T.P. and Ngoc, H.Q. (2001) *Female Labour Migration: Rural-Urban*. Hanoi: Women's Publishing House.

Turner, S. (2009) Informal economies. In: Kitchin, R. and Thrift, N. (eds.), *International Encyclopedia of Human Geography*. Amsterdam: Elsevier.

Turner, S. and Oswin, N. (2015) Itinerant livelihoods: Street vending-scapes and the politics of mobility in upland socialist Vietnam. *Singapore Journal of Tropical Geography*, *36* (3), 394–410.

Turner, S. and Phuong, N.A. (2005) Young entrepreneurs, social capital and Doi Moi in Hanoi, Vietnam. *Urban Studies* 42 (10), 1693–1710.

Turner, S. and Schoenberger, L. (2012) Street vendor livelihoods and everyday politics in Hanoi, Vietnam: The seeds of a diverse economy? *Urban Studies* 49 (5), 1027–1044.

UNDP Viet Nam (1998) Expanding Choices for the Rural Poor: Human Development in Vietnam, Hanoi

Ureta, S. (2008). To move or not to move? Social exclusion, accessibility and daily mobility among the low income population in Santiago, Chile. *Mobilities* 3 (2), 269–289.

Urry, J. (2000) *Mobilities for the Twenty-first Century*. London: Routledge.

Urry, J. (2004) Connections. *Environment and Planning D: Society and Space* 22, 27–37.

Uteng, T.P. (2009). Gender, ethnicity, and constrained mobility: Insights into the resultant social exclusion. *Environment and Planning A* 41 (5), 1055–1071.

Van Blerk, K. (2013) New street geographies: The impact of urban governance on the mobilities of Cape Town's street youth. *Urban Studies* 50, 556–573.

Van den Berg, L.M., van Wijk, M.S. and Van Hoi, P. (2003) The transformations of agriculture and rural life downstream of Hanoi. *Environment and Urbanization* 15, 35–52.

Viet Nam Net (2009) Experts surprised by audacity of proposed Hanoi master plan. Available at: http://ashui.com/mag/english/news/1311-experts-surprised-by-audacity-of-pro posed-hanoi-master-plan.html (accessed on 18 September 2017).

Viet Nam News (2008) *Hanoi's vendors given temporary reprieve. 24 January*. Available at: http://vietnamnews.vn/society/173216/ha-nois-street-vendors-given-temporary-repri eve.html (accessed 18 September 2017).

Waibel, M. (2004) The ancient quarter of Hanoi – A reflection of urban transition processes. *Asien* 92, 30–48.

Waibel, M., Dörnte, C. and Schröder, F. (2011). Hanoi goes West: Stadterweiterung, akteure und konsequenzen. *Geographische Rundschau* 59 (9), 48–54.

Yasmeen, G. and Nirathron, N. (2014) *Vending in Public Space: The Case of Bangkok.* Women in Informal Employment Globalizing and Organizing (WIEGO) Policy Brief No. 16, May. Available at: http://www.wiego.org/sites/default/files/publications/files/Yasmeen-Vending-Public-Space-Bangkok-WIEGO-PB16.pdf (accessed 18 September 2017).

3 Exploring the intersection between physical and virtual mobilities in urban South Africa

Reflections from two youth-centred studies

Gina Porter, Kate Hampshire, Ariane De Lannoy, Nwabisa Gunguluza, Mac Mashiri and Andisiwe Bango

The mobile phone is transforming African mobile lives at a variety of scales, from the minutiae of individual spatial orientations to expansive global connectivities. Now-possible fluid interdependencies between corporeal mobility and virtual mobility have the potential to reframe and reshape lives, especially for young people, who typically have limited financial resources yet are often at the vanguard of mobile phone adoption. This chapter explores the intersection between physical and virtual mobilities among young people in two smaller South African urban centres, drawing on mixed-methods field research, the first study conducted with young people 9 to18 years, the second with a wider age group extending from 9 to 25 years. It focuses on the transport and related physical mobility challenges young people face in reaching locations (and people) important in their lives, and the role that access to mobile phones is now having in mediating those challenges and associated access patterns in these sites. Particular attention is given to the role of gender in the shaping and reshaping of mobility and access patterns: precarity, safety and security are significant themes.

Introduction

Unlike much of the Global North, few homes in poorer (predominantly black) neighbourhoods in South Africa, urban or rural, ever had access to landlines. Consequently, following on from the first appearance of mobile phones in South Africa in the closing decade of the twentieth century, this technology has taken hold rapidly, and with dramatic impact. It is now transforming African mobile lives at a variety of scales, from the minutiae of individual spatial orientations to expansive global connectivities. Now-possible fluid interdependencies between corporeal mobility and virtual mobility have the potential to reframe and reshape lives, especially for young people, who typically have limited financial resources yet are often at the vanguard of mobile phone adoption. This chapter explores the intersection between physical and virtual mobilities among young people in two smaller South African urban centres. It offers reflection on the pace of recent change with particular reference to phone use and *daily* mobilities.

Our research on young people and mobile phones started in 2006–2010 as an unanticipated element in a study focused on young people's daily physical mobility and associated access to transport and services – schools, health centres, markets, leisure activities, etc. – across 24 sites in sub-Saharan Africa.[1] We were working only with 9- to 18-year-olds at that time, but – even so – were amazed at the extent to which young people in the South African urban sites, in particular, were already using (mostly basic, not Internet-enabled) phones regularly to reshape and mediate their physical mobility patterns. Further funding has enabled us to build on this work on mobile phones more widely (including regarding broader impacts of mobile phones on education, inter-generational relations, health-seeking behaviours, livelihoods, etc.), but still with firm reference to linkages with physical mobility (Porter et al., 2015a b; Hampshire et al., 2015; 2017). We have also extended our age range in this second project to include young people aged 19 to 25, thus enabling us to follow up on usage patterns for the cohort of young people we first encountered in 2006 who have now moved into their late teens and twenties. In 2006, hardly any of the young people we interviewed had access to smart phones – the standard phone was a basic model. By 2015, Internet-enabled smart phones were an essential accoutrement of 'cool youth': "Because technology is advancing so I must also move along with it; I cannot use the same type of phone all my life, people will laugh at me" (Gauteng Urban, male, 25y).

The two research sites on which we focus in this chapter are poor, high-density neighbourhoods within smaller urban centres in South Africa: one in Eastern Cape, the other in Gauteng Province. Although Gauteng Province has long been a focus of migration from poor (former homeland) regions like Eastern Cape, there are pockets of deep poverty and it was one such neighourhood that we targeted for our study. The young people whom we have researched in both locations often lead difficult lives, with access to few resources: some reside with an unemployed parent or a pensioner grandparent, or – especially in Eastern Cape – are renting a room in order to attend secondary school in town, having migrated in from a rural area. Among those who have already left school, many spend much of their time just waiting at home[2]: these are unemployed young men and women, some of whom still retain the determination to look for a job, despite the low prospects of finding the kind of well-paid, fulfilling work to which they aspire; the lives of other young women are already substantially constrained by babies and child care. Meanwhile, the street scene is often unwelcoming, even hostile, particularly in Eastern Cape – tainted by the threat of violence to innocent bystanders and travellers, especially at night.

Drawing on a substantial mix of qualitative and quantitative data and associated triangulation, we were able to explore how increasing access to mobile phones in these two urban research sites over the past 10 years has been affecting young mobile lives. The chapter first introduces some relevant literature and the methodology employed in the research; subsequent sections focus on the transport and related physical mobility challenges young people face in reaching locations (and people) important in their lives, and the role that access to mobile phones is now playing in mediating those challenges and associated access patterns. Particular attention is given to the role of gender in the shaping and reshaping of mobilties in this era of extensive and intensive phone communication.

The benefits of mobile phones for travel and distance management: reviewing recent debates

Because this chapter is concerned with the impact of phones on urban patterns of daily mobility and transport usage, it is best set first within the wider debate on that topic. The emergence of Information and Communications Technology (ICT) and, in particular, mobile phones, raised early hopes in Western planning circles that some ease in traffic volumes would occur as a response to the substitution of physical journeys with phone and other ICT communication and associated spatial reorganization of daily mobilities. However, empirical evidence to date in Western cities has not supported that contention. Rather, because phones are portable, there are new opportunities to reschedule on the move (Kwan, 2006; Line et al., 2011; Taipale, 2013). Taipale (2013) describes how, in urban Finland, users have developed a 'virtual reservoir of mobilities' by connecting ICT with urban public transport, and notes that women lead this trend of combining physical mobility with ICT. In addition, as Urry (2012) emphasises, there will remain requirements for continued physical co-presence in many situations, given the need to satisfy social obligations. At least occasional face-to-face is also often required for building and preserving trust and tacit knowledge.

Across sub-Saharan Africa, the conditions within which physical travel takes place are somewhat different. Widespread poverty, irregular transport availability in many locations (including some urban and many peri-urban areas) and potentially hazardous journeys on poor roads in badly maintained vehicles, with added risks of harassment and extortion (from highway robbers, etc.) arguably may weigh more strongly in the balancing of virtual with physical mobility than in the Global North (Porter, 2015). Some business-oriented studies, in particular, have presented evidence of respondents reducing their travel as a result of mobile phone communication, because of the low cost of phone messaging and calls compared with the time costs, financial costs and risks associated with travel (e.g. Jagun et al., 2008; Baro and Endouware, 2013). However, these studies tend to be focused on prolonged long-distance journeys, where hazards are likely to be particularly substantial, rather than on local travel within a relatively confined urban context. By focusing on young people in two poor, high-density urban areas of South Africa in this chapter, we offer further insights into this debate.

Another major strand of literature relevant to the chapter concerns gendered patterns of travel and girls' and women's travel safety in urban areas, both with reference to walking and motorised transport. There is a substantial literature on this issue, extending over many decades (e.g. Lynch and Atkins, 1988), but events in recent years (such as the 2012 gang rape of a young woman on a Delhi bus) have brought wider awareness of the scale of this problem and, in some cities, including Mexico City, Dubai and Dhaka, (contentious) moves to introduce so-called 'pink' services, where women travel separately from men, because of rising complaints by women of sexual harassment on public transport (e.g. Dunckel-Graglia, 2013). In South Africa concerns about women's travel safety are widespread, rape statistics alarming and violent robbery and car-jacking remarkably commonplace.[3] Much of this is arguably intimately bound up with a crisis of masculinity (Jewkes and Abrahams, 2002; Jewkes et al., 2011; Porter, 2013).

Despite these concerns about girls' and women's travel security, the literature on mobile phones has had, as yet, surprisingly little to say about their use in the specific context of safety while travelling (as opposed to safety more broadly interpreted). In the Global North there is more focus on their use by women in 'remote mothering', checking on children elsewhere (e.g. Kwan, 2007), though Ling (2004: 35–55) notes how the mobile phone confers a sense of security that legitimates parents giving children their own phone, while Fyhri et al., (2011) and Nansen et al., (2015) offer studies which refer to the phone as a child's companion device, including during travel. In the Global South, work is similarly sparse, though there are often passing references to the use of phones on the road, as for instance by Horst and Miller (2006: 76) who talk of women taking phones for security purposes when they walk to church at night in Jamaica. More detailed examples include Velghe's (2011) work on mobile phones in a South African township (which suggests a tendency for reduced children's mobility because they can phone friends instead of visiting them, and reduced travel for job search) and Porter's (2016) work on phone usage in rural Malawian and Tanzanian mobility contexts. Our data from South Africa presented herein suggest the need for stronger attention to the role of phones in travel security contexts, particularly in low-income urban areas.

Methodology

This chapter draws on mixed-methods research from two studies, both of which were conducted with the assistance of students from local universities in South Africa. Our original child mobility study conducted in 2006–2010 drew on in-depth interviews with children aged 9 to 18 years, their parents and other key informants (approximately 50 interviews in each of the two South African urban sites) and additional school essays and focus groups with children, followed by a survey of c. 125 children c. 9 to 18 years per site (including 123 in Eastern Cape Urban, 125 in Gauteng Urban). Our household selection in the survey was based on cross-settlement transects, followed by within-household random selection. The questionnaire survey was undertaken towards the end of the qualitative phase, when the majority of the qualitative research had been completed; this allowed us to use information from the qualitative interviews to help shape the survey questions. Because the likely importance of virtual mobility emerged early in the qualitative phase, all age groups were asked about young people's usage of mobile phones and this led on to specific questions in the survey regarding young people's access to phones and patterns of use.

The subsequent youth phones study in 2012–2015 in the same research sites followed a similar pattern of qualitative interviews (c. 50 in each of the two South African urban sites) followed by a survey (with questions informed by the qualitative research). Qualitative interviews with young people in this study were built around their phone stories and call registers; other activities included focus groups with diverse ages, school essays, key informant interviews (including with transport operators) and a small number of life histories with people in their late 20s to mid-30s. The survey procedure replicated that of the earlier project, but with the focus on phone use rather than physical mobility per se. Importantly, the survey component was conducted with a broader age group, extending from 9 years through to 25 years, in order to gain some understanding of intersections between physical and virtual mobility among a cohort where many have been conversant with mobile technology for some years. In the survey, we aimed to cover the same number of children aged 9 to 18 years in each research site as in the earlier project (i.e. c. 125), plus an additional 62/63 young people aged 19 to 25 years per site (i.e. our older cohort, 6 years on). The actual numbers for Eastern Cape Urban were 129 aged 9 to 18 years and 58 aged 19 to 25 years, and for Gauteng Urban, 154 aged 9 to 18 years, and 53 aged 19 to 25 years; the slightly lower than desired coverage of 19- to 25-year-olds reflected the difficulties of finding people in this older group willing to be interviewed in these two urban high-density, high-crime, high-poverty sites.

Here fieldwork was also potentially dangerous for our team, despite positive support from community leaders. It thus required very careful attention to the safety procedures we instituted: interviewers worked close together in nearby compounds, with easy access to a vehicle (in case it should be necessary to vacate the area quickly). In Eastern Cape Urban, where street violence was particularly strongly in evidence, we benefitted from the well-established links between our lead field coordinator and the community.

Contextual information about daily mobility and its hazards (based on data collected 2006–2010)

Walking hazards

The heavy dependence on pedestrian transport in both South African urban centres (as across much of sub-Saharan Africa) was one of the strongest themes of our 2006–2010 child mobility study. This was despite the fact that, in both settlements, almost all those interviewed lived within 15 minutes' walk of public transport (and more than 60% within five minutes' walk).

Walking inevitably dominated as a mobility mode among young people (all c. 9–18 years in this project) in these sites because they lacked funds to pay transport fares. However, walking was widely disliked as a transport mode for numerous reasons, particularly in the Eastern Cape Urban site. Fears of tsotsis, other thugs, thieves and mugging were raised as major concerns in the child mobility survey by more than a quarter of girls here, and nearly a quarter of boys (compared with under one tenth of girls and just over one tenth of boys surveyed in Gauteng Urban): “I don’t like to travel to school because there are boys who mock us on the way to school. They wait for us on the road where they smoke dagga (cannabis) and then they follow”. (Eastern Cape Urban, girl, 13y). A slightly younger girl in this settlement identified

> places we do not dare to go. There is a small shack where these boys smoke their drugs. It is a short cut but I will not use it because these boys can hurt you. . . . There are some places . . . where I have to go with my sister or my friends because of the boys in the settlement. They always want to stop you and propose love to you, so it is better to walk with somebody.

Mocking and other forms of harassment were feared not only in themselves, but because of the danger that they might transform into something worse – in particular, rape. In both Eastern Cape and Gauteng urban settlements, approximately 7% of girls (but no boys) said they specifically feared rape.

Generally, the safest procedure, especially where girls were concerned, was considered to be walking in a group: “We have to walk in groups because there are boys who are not schooling who take our money and mobile phones. If we walk in a group it is better, because some are known by other students” (Eastern Cape Urban, girl, 18y). However, boys also reported fears associated with walking alone, especially after dark and similarly often preferred to walk in a group:

> It is not safe to walk at night alone. You can be mugged by the tsotsis. Even during the day there are boys who smoke dagga and sit at the street corner. If they don’t know you, they will stop you and ask for 2 Rand. If you don’t give them they will slap you on the face.(boy 12y)

One mother observed, “Walking from the taxi rank to this house is a mission, because you don’t know what your child will encounter on the way!” Even walking with an older person did not necessarily offer protection. The potential to encounter violence while walking often appeared to be simply a matter of being in the wrong place, at the wrong time: “When I was walking in our street one day, I saw one boy being shot by some boys who belong to a gang from the other location. . . . I was terrified and ran into one of the houses” (boy, 12y). “Sometimes we see dead people on the way to school and that also disturbs our performance” (girl, 15y).

Motor transport hazards

The most common motor transport used by our respondents living in both the Eastern Cape and Gauteng urban research sites was the minibus taxi, but this was especially the case in Eastern Cape Urban. Here about one third of girls and a quarter of boys surveyed said they travelled by this mode at least once per week, though only about 5% did so on a daily basis. When asked about dangers of travel by motor transport in the survey, the overwhelming emphasis of girls and boys was on fear of traffic accidents; no reference was made to dangers of rape or robbery/thugs. But in the qualitative interviews, by contrast with the survey, girls put emphasis mostly on travel risks associated with harassment and violence: seemingly, it was only within the more informal context of those extended, in-depth conversations that such sensitive elements could rise to the surface. Perhaps the survey context also explains why, in South Africa's 2013 National Household Transport Survey, only 0.2% of those females who reported not travelling in the seven days before the survey indicated that this was due to security risks on transport, despite the fact that information from operators offered a very different picture (Fia Foundation, 2016).

Minibus taxi drivers are a particular source of harassment in South Africa. The following examples of complaints from each of the urban sites are reflective of wider views (similarly found in the peri-urban and rural research sites):

> Sometimes taxi drivers harass us; like if you are a girl and the last to disembark, they want to propose to you and they could just drive around with you and pass your home.
>
> (Gauteng Urban, girl, 17y)

> The thing that I fear about travelling on a minibus taxi is that the drivers propose love to us and they say they want us to be their girlfriends. I am afraid they might kidnap or rape me if I am alone in the taxi . . . the taxi conductors are very rude. Just because we are girls they talk trash and vulgar language to us. They don't have respect. One time I was travelling with my friend from town and when we were disembarking the taxi, the conductor touched my friend's back (buttocks). When we asked him what was his problem, he started talking vulgar language to us and the driver was laughing. (Eastern Cape Urban, girl, 12y)

The scale of harassment is such that one mother in the Eastern Cape urban site observed that no girls should travel by minibus taxi till they were at least 16 years old: "If they are younger they will be raped by taxi drivers".

Minibus taxi drivers' (seemingly well-founded) reputation for preying on girls appears to be reinforced by local meanings of masculinity. These men form very powerful, often feared, collective groups in South Africa; consequently, all ages and both genders tend to be extremely careful in their interactions with them:

> The taxi is far more relaxing than the bus. . . . The disadvantage of travelling by taxi is that drivers are rude and they talk vulgar language. . . . I don't backchat taxi drivers. If they say you must do this you must do it, otherwise they will make you disembark their taxi. (Eastern Cape Urban, boy, 12y)

Nonetheless, as occasional respondents observed, some young girls fall for these (often older) men, attracted to the potential status and benefits that sexual liaison with a taxi driver may bring – and apparently oblivious to the potential dangers (Leclerc-Madlala, 2003: 222–223; Luke, 2003, 2005). One minibus taxi driver (a 40-year-old), observed: "Girls are actually dating these drivers (who) promise them money. It is shocking as you will see a 16-year-old girl having an affair with a taxi driver" (Eastern Cape Urban). Despite these issues, minibus taxi remains the main motor transport mode. Only a few young people had parents, friends or neighbours with a private car: indeed, nearly three quarters of both girls and boys in these settlements had never travelled by private car.[4]

Reflecting on gendered mobility patterns

Mobility and immobility, we have argued elsewhere, are key factors shaping young people's urban experience and their future life chances in Africa (Porter et al., 2010). The data presented previously show that in these two South African urban neighbourhoods young people's physical mobility – especially girls' – is substantially constrained, not least by the threat of harassment, whether they are on foot or travelling in a minibus taxi (the two main transport modes). Additionally, girls may be perceived as not only vulnerable but also potentially promiscuous and thus the focus of substantial parental/family efforts to constrain their mobility. There are fears that boys, meanwhile, may be persuaded onto the streets and into criminality: drugs, theft and gun crime: "I will praise God if my children do not fall into the temptations of this society. Girls are getting pregnant, and boys are stealing and turning to crime" (Eastern Cape Urban, mother of two girls and one boy). However, while we have focused mostly on the down-side of travel for young people in this section, it is also important to recognise the potential that mobility along city streets also offers for excitement, thrills, inclusion and opportunity; not least, opportunities for meeting members of the opposite sex (Porter et al., 2010; 2017).

Mobile phone usage and its impacts on physical mobility in the two urban study neighbourhoods

In both positive and negative mobility arenas, the potential of mobile phones is considerable, whether as an alternative or adjunct to physical mobility, as this section demonstrates.

First, the section charts the expansion of youth ownership and usage of mobile phones in the two research sites and then considers the impact that mobile phone ownership is having on everyday mobile lives.

The expansion of mobile phone ownership and use in the study sites

Even when we conducted our child mobility study in 2006–2010, mobile phones were firmly in evidence in the two urban study sites. However, mobile phone ownership has grown substantially in both urban neighbourhoods (as in South Africa as a whole) over the six years between this and our more recent phone study. Because we only have data for the 9- to 18-year cohort for the earlier child mobility survey, we are limited to a comparison of this age group for the two surveys; even so, the evidence of expansion is impressive (Table 3.1).

In these settlements ***ownership*** looks to have roughly doubled overall (from an average – when calculated as a total of both genders – of 30.3% to 60% in Eastern Cape and from 28% to 61.7% in Gauteng). However, the expansion looks particularly remarkable among females in our sample in Eastern Cape Urban.[5] When we look at phone ***usage*** in the week before the survey (Table 3.2) it is clear, in both surveys, that usage extends well beyond the individual's personally owned phone: rather, we have widespread evidence of sharing and borrowing of phones (among family, friends and neighbours) in both study periods and from both qualitative and survey data. Nonetheless, a significant expansion in usage has occurred for both males and females between the two surveys, especially among males in Gauteng Urban (where usage was substantially lower than among females in 2007–2008) and females in Eastern Cape Urban. In Gauteng, three quarters of 9- to 18-year-olds of both genders had used a phone in the week before the 2013–2014 survey, more than 90% in Eastern Cape.

The 2013–2014 survey allows us to extend observation to young people aged 19 to 25 years (Table 3.3), many of whom are likely to have been regular phone users for some years. Here, the total sample size is relatively small, but ***ownership*** looks remarkably high, especially among women.[6]

Table 3.1 Percentage of young people 9 to 18 years old with own phone (all types), 2007–2008 and 2013–2014

	2007–2008 (N = 248)		*2013–2014 (N = 285)*	
	Female	Male	Female	Male
Eastern Cape Urban	29.8%	32.4%	70.3%	50.8%
Gauteng Urban	27.0%	29.4%	57.7%	65.1%

Table 3.2 Percentage of young people 9 to 18 years old who had used a mobile phone in the week before survey, 2007–2008 and 2013–2014

	2007–2008 (N = 248)		*2013–2014 (N = 285)*	
	Female	Male	Female	Male
Eastern Cape Urban	73.2%	78.4%	98.5%	90.8%
Gauteng Urban	69.4%	46.9%	74.7%	75.9%

Table 3.3 Percentage of young people 9 to 18 years old and 19 to 25 years old currently owning at least one phone in working order (all types), 2013–2014

	9–18y (N = 285)		*19–25y (N = 111)*	
	Female	Male	Female	Male
Eastern Cape Urban	70.3%	50.8%	90.9 %	76.0%
Gauteng Urban	57.7%	65.1%	100%	84.8%

Table 3.4 Percentage of young people 9 to 18 years old and 19 to 25 years old who had used a mobile phone in the week before survey, 2013–2014

	9–18y (N = 285)		*19–25y (N = 111)*	
	Female	Male	Female	Male
Eastern Cape Urban	98.5%	90.8%	100	96.0
Gauteng Urban	74.7%	75.9%	100	87.9

Usage of phones among 19- to 25-year-olds is, unsurprisingly, even higher than among 9- to 18-year-olds (Table 3.4). In the case of women aged 19 to 25, in both sites all had used the phone not merely in the past week but on the day of the survey or the day before. As the qualitative data confirm, for these women the mobile phone is critical to the conduct of daily life.

Implications of expanded mobile phone ownership and use for mobility patterns

So, what does all of this mobile phone expansion mean for young people's physical mobility and transport usage? Our preliminary research on this theme, based on the 2006–2010 data set, suggested that in the South African urban sites it was girls, in particular, who were already starting to use mobile phones to re-envision their mobility opportunities. The virtual mobility offered by the mobile phone (for planning journeys, organising clandestine meetings, assessing destination potential) was starting to present a particularly potent tool in the repertoire of obfuscation and circumvention which young people needed to employ when their mobility was constrained by adult restrictions (Porter et al., 2010; Porter et al., 2017). We also observed young people starting to substitute phone communication, on occasions, for physical travel, and using the phone to assist in organisation of transport and travel, including in emergency contexts. We did not capture specific data on this theme in the survey, but even at this early stage of mobile phone uptake, qualitative research offered many stories of lives saved because phone communication enabled timely access to transport, often with reference to obstetric emergencies.

Our latest 2012–2015 study, with its specific focus on the impacts of mobile phones on young lives, has enabled us to consider the implications of virtual mobility in much greater depth than was feasible in our first (2006–2010) research phase. We now included questions about perceived impacts on travel practices in the survey, while the addition of the older cohort (19–25 years) also enabled some wider perspectives, including from young people who have left school and are pursuing livelihood activities or searching for work. The following three subsections cover, firstly, how phones are used to help deal with travel difficulties; secondly, respondent perceptions of mobile phones' impact on local and long-distance travel; and thirdly, implications for travel safety and security.

Impacts of mobile phone usage on dealing with travel difficulties: In the 2012–2013 survey, we asked questions about the use of mobile phones in the previous

12 months with regard to responding to transport difficulties. In Eastern Cape Urban, 41% of females and 44% of males aged 9 to 25 years had used a mobile phone for such purposes, with slightly lower proportions (34% of females, 39% of males) in Gauteng Urban. The main transport difficulties which required phone communication in both these locations tended to centre round no transport being available or not arriving when required; occasionally it related to traffic jams. Qualitative data supported these findings, albeit mostly mirroring earlier observations from the 2006–2010 child mobility study about the value of phones for organising transport in health emergencies:

> My sister went to visit my aunt in (township) and when she came back she wasn't feeling okay. Late at night she started to sweat and she had a headache. My mother wasn't home. I called an ambulance to come and pick her [up]. It came and took her to the hospital. I called my mother to inform her that I was in the hospital. (Gauteng Urban, male, 26y)

Impacts of mobile phones on local and long-distance travel: In the context of debates around the impact of mobile phones on physical travel in the Global North, we were particularly keen to discover whether mobile phones were affecting overall travel, both local and long distance. In the survey, we thus asked respondents of all ages (i.e. 9–25 years) who had used a mobile phone in the past 12 months for communication purposes (thus, not merely use for games or as a calculator, etc.) about the perceived impact of mobile phone usage on their short day-to-day local journeys and irregular long-distance journeys (Tables 3.5 and 3.6).

So far as short and long journeys are concerned, both sites show more reduction than expansion in journeys associated with phone use, which makes sense in the context of young people's usually constrained financial resources and prevailing

Table 3.5 Perceived impact of phone use on short, day-to-day local journeys, 2013–2014

	Eastern Cape Urban		*Gauteng Urban*	
	Male %	Female %	Male %	Female %
No impact	13.4	14.1	67.4	72.2
More journeys	22.0	26.1	15.7	8.3
Fewer journeys	64.6	59.8	16.9	19.4

Table 3.6 Perceived impact of phone use on long, irregular journeys, 2013–2014

	Eastern Cape Urban		*Gauteng Urban*	
	Male %	Female %	Male %	Female %
No impact	27.2	29.2	63.1	72.2
More journeys	27.2	23.6	9.5	8.3
Fewer journeys	45.7	47.2	27.4	19.4

high motor transport fares. Cheap calls/texts (which can be built round network operators' special promotions such as Vodacom's nightshift, MTN's Mahala Thursday, free 'call backs' and contact through Mxit or WhatsApp) were widely assessed as a satisfactory substitute on many occasions for physical travel.[7] As one young man succinctly explained, with reference to collecting examination results: "If you have R2 airtime you just SMS your student number. . . . Then your results will be sent you, rather than spending a lot of money for transport" (Eastern Cape Urban, 22y). Phone communication was also recognised as supporting more efficient travel, because it avoids wasted journeys. It is possible to check first to ensure the person to be visited will be at home – and now that most adults in these urban sites have easy access to a mobile phone, this is standard practice.

Unsurprisingly, a majority of both genders in both research sites assessed these perceived travel reductions as positive (particularly in Eastern Cape Urban); only one fifth or less across both sites saw the perceived changes as negative. In some cases the phone reduced everyday short journeys such as saying hello to a schoolmate or organising child care; in others it supported connections otherwise rarely feasible, as with family members living in distant parts of the country or people unwelcome in the family home (often boyfriends, but also a number of estranged parents). Virtual communication thus appeared to be widely supporting both intra-familial, inter-generational linkages and intra- and extra-familial linkages with people the same age (Porter et al., 2015b).

However, as Tables 3.5 and 3.6 indicate, there is a very distinctive and intriguing difference between the two sites regarding perceived overall impact of phone use on travel, which is assessed as far more substantial by Eastern Cape Urban respondents than by those in Gauteng Urban (male and female). This may be related to two factors, firstly the lower usage of phones overall (except among women aged 19–25 years) in Gauteng Urban, which will somewhat reduce the potential to connect by phone to the wider population of contacts, and secondly – and likely to be far more significant – the fact that physical mobility in the Eastern Cape Urban neighbourhood appears to be particularly hazardous for both males and females (as both qualitative and survey data for 2006–2010, discussed earlier, indicates). This may well explain the very substantial perceived reduction in short journeys in Eastern Cape and the fact that this applies to both males and females. This is discussed further in the next section.[8]

Mobile phones and travel security: Qualitative data from the 2012–2015 study not only adds texture to the story of mobile phone impacts on physical mobility but also links particularly to issues of travel safety and security. Indeed, in Eastern Cape Urban, where life on the street is extremely hazardous for both genders, the potential for phone theft itself (already becoming evident in the 2006–2010 period, see Porter et al., 2010) adds substantially to travel dangers, especially when on foot:

> Phone thefts and attacks is a common thing here in both males and females. They (robbers) look at you; if you are not familiar to them, they attack you.

> Even when you are physically weak or drunk they attack you. They do it almost all the time. In the morning, they target students and people who are going to work (commuters); the same applies in the afternoon and evening. Their hot spots are between (this township and one nearby). There are many taverns in that vicinity so they stay there and wait for their victims. Sometimes police come and patrol in that vicinity but not all the time. Some of these thugs carry knives, others guns. I'm not sure whether those guns are legal or not. (male, 23y)

Thefts of phones and other valuables were also reported in Gauteng Urban. For instance, one young woman aged 17 (whose mother now calls her regularly to check she is safe when she is out on the streets) was just returning home from a local store with her sister one evening when she was attacked at gunpoint by thieves who made off with her cell phone. But these events seem to be rarer than in Eastern Cape Urban.

While possession of a mobile phone can add to young people's vulnerability to attack on the street, for young women in Eastern Cape Urban, in particular, there is specific evidence in the 2012–2015 study of the positive security role that mobile phones now play in terms of organising transport. One 18-year-old secondary school girl explained how, if they are late finishing at school, they can call their principal who will then take them home by car, 'since it's not safe here in our township'. Another recounted how she used to call a taxi to travel home from school in the evening even over very short (walking) distances 'because of the robbers'.

The mobile phone also presents opportunities to ease travel safety and security through virtual escorting and way-finding, as in the following examples. In Gauteng Urban, a 12-year-old girl recounted how she had become separated from her sister in the city centre one day:

> (suddenly), she was not there. All I could see was just a lot of strangers who were just busy walking around, I was so scared and I wanted to cry. I had my mother's phone with me so I sent a 'Please call me' message because the phone had no airtime. She called me immediately and asked me where I was. She told me to stand (there and) she came and took me.

Another girl, around the same age, who lives with her grandparents, explained how phones came into her travel story when she had taken a taxi home after a visit to her mother in a peri-urban location. Immediately after reaching the home taxi stop, she sent her mother a 'personalised call me "arrived" ', but then her mother called her back to ensure not only that she was safe but also that she was still in possession of the cash[9] which had been sent with her for her grandparents.

Many urban taxi drivers told us they now give passengers their mobile number as a matter of course, in the hope that this will lead to regular customers. At night, when dangers are especially great, young women often now report using their cell phones to arrange for private taxicabs to take them home, although fares are expensive:

> I use my cell phone to also arrange transport like calling a cab to come and take me, especially when I came late from work. I just call XXX, a cab driver, to come and pick me [up]. The advantage of using a cab – it's safe, like it drops me here in the yard, unlike a taxi that will drop me on the road (where) I become vulnerable to thugs. So, it's safer to use a cab. Even in town you know it comes straight to my work place, so I don't have to move around the streets and run away from thugs. But the most disadvantage of it is that it's (costly) – hey, like it cost R60 from town to here whilst in a (minibus)taxi it's only R5 but that does not happen on daily basis. (Eastern Cape Urban, female, 20y)

Conclusion: mobile phones and changing mobilities/ transport landscapes

Mobile phones already look to be fulfilling some of their evident potential for supporting distance management across the world, both in emergency and everyday travel contexts, and in urban as well as rural settings. However, whereas in the Global North mobile phone usage appears to have done little, as yet, to improve traffic congestion and the overall reductions in motorised transport that are essential for reduced carbon emissions, in the two South African urban study sites discussed in this chapter, the evidence for positive impact looks, in some respects at least, more encouraging.

Young people's physical mobility practices in these sites are substantially shaped not only by low disposable incomes but also by fear. The widespread availability of mobile phones now interposes opportunities to both improve personal safety and to make better use of their limited funds for distance management, by substituting virtual for physical travel whenever feasible. Whereas safety appears to be the principal factor behind the substitution of phone communication for many short, daily (often pedestrian) journeys of both genders on the streets (especially in Eastern Cape Urban), financial considerations come more strongly into play when it comes to contemplating long (expensive) journeys by motor vehicle (though safety may still figure where the phone is brought to bear as a virtual escort or way-finder).

Taken together, the perceived reductions in both long-distance, irregular and short, everyday journeys appear to be quite substantial. However, these reflections are based on our respondents' stories and perceptions of change: the extent to which they translate into lower than might be anticipated (pedestrian and motorised) traffic flows on the ground in these low-income neighbourhoods needs further investigation. And, of course, if there are reduced numbers of people walking on the streets, what does this mean in the longer term for health (obesity levels) and security? Arguably, the presence of fewer people will actually increase the dangers of pedestrian travel, thus encouraging reliance on motorised transport for essential journeys: potentially a vicious, not virtuous, circle! A related question concerns the extent to which substitution of phone communication for transport may also be occurring in those high-density urban neighbourhoods in the Global North where precarity and fear of violence similarly prevail widely. Is widespread mobile phone usage perhaps encouraging substitution of virtual for physical mobility here too, in the absence of safe, good-quality public transport? If so, what form does this take? Is growing immobility occasioned by precarity and fear of walking in urban neighbourhoods perhaps one element contributing to the growing prevalence of obesity among poorer populations? Of course, however, any impact in terms of motor traffic volumes in both Global South and North may well be obscured by the mobility practices of the richer population who dominate private vehicle ownership.

Notes

1 Both studies were funded by the UK Economic and Social Research Council and the Department for International Development (ES/D002745/1; ES/J018082/1). The child mobility team 2006–2010 for South Africa included all authors except Ariane De Lannoy and Nwabisa Gunguluza, who joined our team at the start of the second project.

2 Only 17% of young people no longer enrolled in any kind of educational programme that we surveyed in the Gauteng Urban site had had *any* employment in the 12 months before the survey, whereas virtually all (96%) no longer enrolled in the Eastern Cape urban site had had at least a short phase of employment in that period (though this appears to have often been through temporary return to help out in rural homes).

3 https://africacheck.org/factsheets/factsheet-south-africas-201516-crime-statistics/ This draws on South African Police Service data which notes 42,596 rapes in 2015–2016, but substantial under-reporting is likely. On average 363.1 robberies with aggravating circumstances (using a gun or weapon) were recorded each day. On average 40 cars were hijacked per day; 50% of the car-jacking crime occurred in Gauteng.

4 Bicycles are also rarely a key transport mode. Around four fifths of girls in both sites said they had never cycled (though nearly half of boys in Eastern Cape Urban cycled occasionally) (see Porter et al. 2017, Chapter 7).

5 In 2013–2014, the figures specifically refer to phones currently owned that are *in working order* whereas we asked simply about current ownership in 2007–2008 so expansion may be even greater.

6 Parents often buy their children phones, so they can keep in touch, but from adolescence onwards, parents and others (as in our 2006–2010 study; see Porter et al. 2012) regularly express concerns about the source of girls' high-quality phones. Older men are frequently accused of using these as a lure for sexual favours (Porter et al. 2012): *sugar daddies. . . . They buy them expensive phones, clothes and jewellery* (girl, 19y).

7 Though there were also cases of schoolchildren buying airtime from the transport money they had been given and walking to school instead.

8 Interestingly, when data for all eight South African sites are combined (i.e. including peri-urban, rural and remote rural sites), females are more likely than males to perceive reductions in their long-distance journeys. This has led us to suggest that major safety considerations may encourage females, in particular, to reduce long journeys when they are able to do so. The isolation of data for poor, high-density, urban areas suggests male vulnerability to attack may be higher here than elsewhere.

9 Dangers associated with carrying money on journeys have encouraged the expansion of mobile phone–based banking services such as e-wallet. However, under one tenth of each gender surveyed (in both settlements) said they had sent money by phone using

mobile money services over the previous 12 months, while only a few additional percent had received money by phone (still well under one fifth; the maximum recorded was 15.5% of males in Eastern Cape Urban). Because many adult South Africans already have bank accounts, bank transfers are a more common mode of moving larger sums of money (including by mobile phone), while small sums are regularly transferred simply by sending airtime.

References

Baro, E. and Endouware, B. (2013) The effects of mobile phone on the socio-economic life of the rural dwellers in the Niger Delta region of Nigeria. *Information Technology for Development* 19 (3), 249–263.

Dunckel-Graglia, A. (2013) 'Pink transportation' in Mexico City: Reclaiming urban space through collective action against gender-based violence. *Gender and Development*, 21 2, 265–276.

FiA Foundation (2016). *Safe and Sound: International Research on Women's Personal Safety on Public Transport*. Available at: www.fiafoundation.org/blog/2016/march/safe-and-sound-the-challenge-of-ensuring-a-fair-transport-system.

Fyhri, A., Hjorthol, R., Mackett, R.L., Fotel, T. and Kyttä, M. (2011) Children's active travel and independent mobility in four countries: Development, social contributing trends and measures. *Transport Policy* 18 (5), 703–710.

Hampshire, K.R., Porter, G., Owusu, S.A. et al. (2015) Informal m-health: How are young people using mobile phones to bridge healthcare gaps in Sub-Saharan Africa? *Social Science and Medicine* 142, 90–99.

Hampshire, K., Porter, G., Mariwah, S., Munthali, A., Robson, E., Owusu, S., Abane, A. and Milner, J. (2017) Who bears the cost of 'informal health'? Health-workers' mobile phone practices and associated political-moral economies of care in Ghana and Malawi. *Health Policy and Planning* 32 (1), 34–42.

Horst, H. and Miller, D. (2006) *The Cell Phone: An Anthropology of Communication*. Oxford: Berg.

Jagun, A., Heeks, R. and Whalley, J. (2008) The impact of mobile telephony on developing country micro-enterprise: A Nigerian case study. *Information Technologies and International Development* 4 (4), 47–65.

Jewkes, R. and Abrahams, N. (2002) The epidemiology of rape and sexual coercion in South Africa: An overview. *Social Science and Medicine* 55 (7), 1231–1244.

Jewkes, R., Sikweyiya, Y., Morrell, R. and Dunkle, K. (2011) Gender inequitable masculinity and sexual entitlement in rape perpetration South Africa: Findings of a cross-sectional study. *PloS one* 6 (12), p.e29590.

Kwan, M-P. (2006) Transport geography in the age of mobile communications. *Journal of Transport Geography* 14, 384–385.

Kwan, M-P. (2007) Mobile communications, social networks, and Urban travel: Hypertext as a new metaphor for conceptualizing spatial interaction. *The Professional Geographer* 59 (4), 434–446.

Leclerc-Madlala, S. (2003) Transactional sex and the pursuit of modernity. *Social dynamics* 29 (2), 213–233.

Line, T., Jain, J. and Lyons, G. (2011) The role of ICTs in everyday mobile lives. *Journal of Transport Geography* 19 (6), 1490–1499.

Ling, R. (2004) *The Mobile Connection: The Cell Phone's Impact on Society*. Morgan Kaufmann.

Luke, N. (2003) Age and economic asymmetries in the sexual relationships of adolescent girls in Sub-Saharan Africa. *Studies in Family Planning* 34 (2), 67–86.

Luke, N. (2005) Confronting the 'sugar daddy' stereotype: Age and economic asymmetries and risky sexual behaviour in urban Kenya. *International Family Planning Perspectives* 31 (1), 6–14.

Lynch, G. and Atkins, S. (1988) The influence of personal security fears on women's travel patterns. *Transportation* 15 (3), 257–277.

Nansen, B., Gibbs, L., MacDougall, C., Vetere, F., Ross, N.J. and McKendrick, J. (2015) Children's interdependent mobility: Compositions, collaborations and compromises, *Children's Geographies* 13 (4), 467–481. doi: 10.1080/14733285.2014.887813.

Porter, A. (2013) 'What is constructed can be transformed': Masculinities in post-conflict societies in Africa. *International Peace keeping* 20 (4), 486–506.

Porter, G. (2015) Mobile phones, mobility practices and transport organisation in sub-Saharan Africa. *Mobility in History* 6, 81–88.

Porter, G. (2016) Mobilities in rural Africa: New connections, new challenges. *Annals of the American Association of Geographers* 106 (2), 434–441.

Porter, G., Hampshire, K., Abane, A., et al. (2010) Moving young lives: Mobility, immobility and inter-generational tensions in urban Africa. *Geoforum* 41, 796–804.

Porter, G., Hampshire, K., Abane, A., et al. (2012) Youth, mobility and mobile phones in Africa: Findings from a three-country study. *Journal of Information Technology for Development* 18 (2), 145–162.

Porter, G., Hampshire, K., Abane, A., Munthali, A., Robson, E., Bango, A., de Lannoy, A., Gunguluza, N., Tanle, A., Owusu, S. and Milner, J. (2015b) Intergenerational relations and the power of the cell phone: Perspectives on young people's phone usage in sub-Saharan Africa. *Geoforum* 64, 37–46.

Porter, G., with Hampshire, K., Abane, A., Munthali, A., Robson, E. and Mashiri, M. (2017) *Young People's Daily Mobilities in Sub-Saharan Africa: Moving Young Lives*. London: Palgrave Macmillan.

Porter, G., Hampshire, K., Milner, J., Munthali, A., Robson, E., de Lannoy, A., Bango, A., Gunguluza, N., Mashiri, M., Tanle, A. and Abane, A. (2015a). Mobile phones and education in sub-Saharan Africa: from youth practice to public policy. *Journal of International Development* 28, 22–39.

Taipale, S. (2013) The dimensions of mobilities: The spatial relationships between corporeal and digital mobilities. *Social Science Research* 43, 157–167.

Urry, J. (2012) Social networks, mobile lives and social inequalities. *Journal of Transport Geography* 21, 24–30.

Velghe, F. (2011) *Deprivation, Distance and Connectivity: The Adaptation of Mobile Phone Use to a Life in Wesbank, a Post-Apartheid Township in South Africa*. Available at: www.tilburguniversity.edu/upload/cdcdd501-1a7b-4af2-803d-dd51dd6623cd_tpcs paper (last accessed 9 January 2017).

4 Informal mobilities and elusive subjects

Researching urban transport in the Global South

Jennifer O'Brien and James Evans

Unregulated motorcycle transport is vital to billions living with poor road access, yet it is increasingly marginalised in transport policies intended to modernise cities across the Global South. This chapter draws on experiences conducting a project that investigated how the estimated 145,000 boda boda motorcycle taxis in Kampala, the rapidly growing capital of Uganda, provide mobility and income for its poorest inhabitants. To capture these moving targets we designed a project using GPS and video technology to capture the movements and uses of boda bodas in more detailed ways than has previously been managed. The chapter focuses on how the research was shaped by the messiness of the field and the role that the boda boda drivers assumed as co-producers of both data and project goals. Adopting a more autobiographical voice, we show how the city shaped our research and draw tentative lessons for other researchers of mobilities in development contexts. In particular, we highlight the ways in which technology can enable co-production and suggest that new approaches to informal transport and mobility are urgently required in the face of rapid urban growth.

Introduction

The experience of being in cities across the Global South is defined by largely unregulated and informal forms of transport. Rickshaws, tuk-tuks, jeepneys, minibuses and motorbikes appear in all sorts of motorised and non-motorised forms across Asia, Africa and South America, permeating and defining the experience of these cities (Simone, 2011). Unregulated and unplanned informal transit accounts for 80 to 90% of public transport journeys in medium-sized cities in the Global South (Luthra, 2006), and in the absence of planned transport infrastructures, it has grown at the same breakneck speed as cities themselves (Cervero and Golub, 2007). Mobility is a key driver of economic and social development because it determines access to jobs, goods and services (UN Habitat, 2010). As a research team working on minibus transport in Nairobi recently suggested, transit is so essential to urban life that people will find a way to develop it with or without government help (DigitalMatatus, 2014; Williams et al., 2015).

Despite its prevalence, very little is known about informal transport and the role it plays in sustainable urban development and poverty alleviation. Worse still, informal transport solutions are increasingly being marginalised in planning policies intended to modernise cities across the Global South, viewed as an unruly challenge to be overcome. This chapter describes our experiences using a combination of Global Positioning System (GPS) tracking and qualitative methods to investigate how the estimated 145,000 boda boda motorcycle taxis in Kampala, the rapidly growing capital of Uganda, provide mobility and income for its poorest inhabitants. The research was inspired by a visit to Uganda in the summer of 2014, which involved a brief stay in Kampala before a longer visit upcountry to visit Jennifer's PhD field-site in Kibaale. We were both struck by the chronic congestion in Kampala and the army of motorcycle taxis that existed to circumvent it. Upon our return we were surprised to find relatively little literature addressing the topic. To some extent this reflects the difficulties of studying informal transport, which operates at the fringes of the institutionalised system and is often weakly governed by illegal operators or self-formed organisations (Behrens et al., 2016). This chapter explores our experiences addressing this research gap around the role of unregulated informal transport in poverty alleviation and sustainable urban development.

Drawing upon our experiences of ethnographical research in Africa and mobile methods respectively, we conceived a project that we called 'boda mobilities' to use qualitative GPS to reveal how informal motorcycle taxis permeate the city and the routes that they are able to navigate, in order to build a picture of how they support the daily lives of Kampalans. Boda bodas are literally a moving target, and we planned to draw heavily on GPS and video technology to capture the movements and uses of boda bodas in more detailed ways than has previously been managed. In addition to providing valuable information to the city planning authority and other local stakeholders, the project was motivated by a desire to understand how uniquely adapted local transport solutions can contribute to the resilience of those living in the planet's most rapidly growing cities. In this sense, the project adopted the spirit of work in the mobilities paradigm in focusing on the 'experiential fact of moving' (Cresswell, 2011; 556), in order to explore what a broader research agenda for the study of informal mobility might comprise. This programme of work was funded by the National Geographic Society and involved two intensive visits to Kampala in 2015 and 2016.

This was the project that was funded, but in practice things unfolded in interesting and often unforeseen ways. The difficulties of accessing the boda drivers who we wanted to study challenged us to reflect upon both our and their positionality. In particular, technology became a bridge that broke down barriers between us and the socio-economically diverse and hard-to-access population that comprises boda boda drivers. The perpetual traffic congestion, referred to locally as 'The Jam', and equatorial cloud bursts constantly undermined any attempt to maintain a classic detached research position and led to an ongoing renegotiation of the methodological approach. As we found ourselves dependent upon the same forms of informal transport we were studying to get around the gridlocked city, the constraints of traffic and mobility came to shape what we were practically able to do as researchers in the field. The informality that characterized movement in the city guided the research in important ways, and ultimately led us to adopt a less formal approach. In this chapter we recount our research experience in an open and personal manner, highlighting the challenges that we encountered studying a moving and often elusive research topic. It is motivated by a desire to share our changed views on the role of technology in this kind of research and our positionality in relation to an informal infrastructure as much as outlining any particular methodological innovation. This is done in an effort to be useful to others undertaking research amidst the messy chaos of informal transport infrastructure in the Global South.

Boda boda! Motorbikes in the city

The arrival into Uganda is deceptive. As the aircraft sweeps over the pristine blue waters of Lake Victoria where fishermen paddle their dugout canoes from the Ssese Islands, to land into the lush green of Entebbe, the traveller is lulled into a complete false sense of security of the urban hustle and bustle to come. After navigating the long immigration queues and waiting patiently behind the usual United Nations envoy, our encounter with boda bodas, which are normally prohibited from restricted airport space, came surprisingly early. We had landed during the presidential election in which President Yoweri Museveni's 25-year reign was being challenged, again. Entebbe Road that connects the airport to Kampala was at a standstill. Ahead was a fleet of minibuses, blaring music from a makeshift loud speaker on the roof, their windows rolled down whilst campaigners of the main opposition, Besigye, precariously leant out screaming and waving flags. The vehicles were surrounded by approximately 100 boda bodas decorated with lengthy green banana fronds that celebrated the passage of the vehicles. The motorbikes were driven by young men, mostly wearing jeans and black T-shirts, some exhibiting their prowess by riding facing backwards, or even while performing handstands. 'Wow', I said, naively to the friend who had met us at the airport, 'it's great to see that boda drivers get involved in politics'. 'Nah,' replied our friend, 'bodas will do anything for five thousand [Ugandan dollars]'.

Originating as a way to cross the Kenyan-Ugandan border in the 1960s, boda boda (which means border border in Swahili) are motorcycle taxis which have since spread through East Africa as a universal, affordable and flexible form of transport (Sietchiping et al., 2012). Kampala is an ideal place to study the role of informal mobility. The city has grown from 137,000 inhabitants in 1960 to 1.8 million today, a number that swells to 3 million during the daytime. The explosive growth of the city and movements of the vast population to the sprawling central business district constructs chronic congestion to the point of gridlock daily. The daily traffic trauma is not aided by the physical geography of Kampala, with the city being framed by seven hills which dictate only four major arterial routes into the capital.

Despite almost permanent gridlock, the city currently has relatively low car ownership at 15 per 100 (KCCA, 2012; KCCA, 2013). Kampala has few macadamised roads and does not maintain an integrated public transport system. The Kampala Capital City Authority (KCCA) has stated a plan to establish a Bus Rapid Transit (BRT) system on nine main lines, with minibuses and other non-motorised transport infrastructure feeding into it. It has also explored the possibility of using cable cars to better connect the city's hills. Inevitably, funding remains the significant stumbling block to implementation. Beyond these lofty plans, there are four main public transport modes in Kampala: buses, *Matatu* minibus taxis (locally known as taxi), boda boda motorcycle taxis, and 'special hire', which are private taxi cars. Although there are a few studies of transport in Kampala (Kamuhanda and Schmidt, 2009), little has been published specifically on boda bodas (although see Howe and Davis, 2002; Kisaalita and Sentongo-Kibalama, 2007). Kampala's unregulated army of motorcycle taxis dodge and weave through the congested streets and alleys with passengers and cargo riding directly behind the driver. Developed as an organic solution to mobility, boda bodas are uniquely suited to the physical form of the city. In addition to picking up fares on their travels, boda bodas congregate at 'stages' across the city. Stages are like bus stops to which drivers are formally registered under a stage chairman, and provide the main place where they wait for fares. A massive fleet of boda bodas now defines the urban and social fabric of Kampala, the capital of Uganda, serving a socio-economically diverse population, representing a key employment destination for poor populations arriving in Kampala and (most importantly) the only efficient way to navigate the winding, gridlocked streets of Kampala.

As an unregulated, needs-driven and organic solution, boda bodas represent a challenge to the traditional Ugandan understandings of modernity and development. They currently sit at the centre of a Ugandan road safety crisis that has drawn comparison with HIV in terms of its national importance (Diaz Olvera, 2012; Magoola, 2013). As an obvious employment destination for young men in Kampala, boda boda drivers are quickly constructed as untrustworthy and largely vilified as dangerous criminal gangs. At the same time the number of providers has led to a glut of motorbikes on the road, prompting a plague of accidents. Fuelled by public protestation at the unregulated growth and perceived threat of boda bodas, the efforts by the KCCA to bring the unruly growth of boda bodas under control have led to a long-running dispute between the boda boda operators and the authorities over perceived attempts to cleanse the city of their presence (Goodfellow and Titeca, 2012; Red Pepper, 2013).

How this story ends will have a major bearing on the sustainability and development of Kampala in the future. Approximately 70% of the population is under 25 years old and the government of Uganda (2005) anticipates continuous growth to 3.6 million inhabitants by 2020 in the Greater Kampala Metropolitan Area. With in-migration and population growth the city is projected to approach 20 million people by 2040 (KCCA, 2012). The ways in which these struggles play out are significant as the majority of the remaining 4 billion people who the UN estimates will inhabit the planet by 2050 will be added to cities in the Global South. While automobile-dominated transport systems are relatively fixed in the developed world, cities in the Global South have the opportunity to adopt more environmentally and socially sustainable strategies (Swilling et al., 2013). Whilst being essential to social, economic and physical access in the city, very little is known about the movements of boda bodas and the kinds of mobility they enable. Two weeks of highly intensive research was undertaken in Kampala, in December 2015. The research aimed to move away from the traditional understandings of infrastructure as a physical exoskeleton that is removed from reality, and instead to investigate it as an embedded set of spatial practices that respond to the need for motorised transport in the face of chronic resource constraints.

Co-producing a research project (by accident)

We had developed and submitted traditional-style research proposals to three different funding bodies before finally being awarded a grant to support the work. This section describes what happened when our carefully laid research plans encountered the reality of boda bodas in Kampala. Before going into the field, we held a series of Skype meetings with a local organisation called Tugende, who are a social enterprise that provides a loan-purchase scheme for boda boda drivers to enable them to purchase their own motorbikes. Tugende ended up providing contacts with drivers, co-producing and administering the driver survey, and providing GPS tracks from their bike security tracking database as well as data from 1,050 driver application forms for analysis, but this was not our original research plan.

On the first morning in Kampala we were woken up by the sound of traffic as the roar of engines and cacophony of horns floated across the valley to our accommodation. These sounds were omnipresent for the two weeks we spent in Kampala on that trip, and 'The Jam' took on a kind of monstrous presence with a life of its own. The New Vision, Kampala's major newspaper, now publishes hourly traffic updates through its social media about 'The Jam' during rush hour, which in recent years has extended from 6am to 10am in the morning and in the afternoon starts at 3pm running through until at least 8 o'clock in the evening. Our first job of the day was a meeting at the Tugende offices where we had been invited to present the project to the staff to get their feedback. The 15-minute walk to the offices presented an early opportunity to experience a Kampalan main road, which was characterised by a lack of pavements with steep mud ditches to the side. The Tugende compound was easily identifiable by the Boxer motorbikes parked in front of the main office, and the gaggle of drivers waiting around to see various staff members.

Upon our arrival we were met by Lucy, their information officer with whom we had been in contact over the previous six months, and were ushered into a small outbuilding towards the rear of the compound. About 15 staff slowly filed in and arranged themselves on plastic garden chairs for the talk. We gave our pitch and handed out draft copies of the driver survey for comment, which generated a wealth of feedback. It became quickly apparent to us that we had little understanding of boda boda drivers as the staff made suggestions about possible ways to rephrase questions to make them more intelligible and relate more closely to relevant issues. One of these included the safety of the drivers themselves. While habitually presented as a source of danger, a major concern for drivers is their own safety. The practice of assaulting drivers and stealing their motorbikes was prevalent enough to have a specific Lugandan term to describe it. 'Iron-bar man' refers to the practice of hitting a driver over the head with a crowbar before making off with the bike. Figure 4.1 shows a stage we visited on our second day, on the outskirts of the city. Two of the stage members had been assaulted and robbed, one of them twice. Less serious but more common hazards included fare dodgers, diluted fuel, dust and corrupt police.

The discussion opened our eyes to what has subsequently become perhaps the most important element of the work, that boda boda drivers are a completely alienated group within Kampala. Conversations with drivers and quotes from the driver surveys revealed a huge feeling of under-appreciation. Drivers spoke about being proud of the services that they provide and their sense of duty to their customers. They also spoke about their love for the industry, which pays relatively well. In many cases, a single bike was supporting extended families both in Kampala and 'upcountry' (in the rural areas of Uganda). For this reason drivers often spoke about 'eating from their bikes' and the subsequent care they took over the source of their livelihoods. In contrast, boda boda drivers are largely represented by the authorities as undesirable criminals who are responsible for creating traffic chaos, creating a massive sense of disillusionment and hostility towards the authorities. The goal of the boda mobility project subtly shifted from identifying the role of boda bodas in providing mobility towards helping boda boda drivers

Figure 4.1 'How will this help us? I don't understand how you asking us questions will help us.'

Source: Author's photo near Gayaza with Mr Wize, 07/12/2015.

put their case more compellingly, to give them some kind of voice in the debates around planning and transport in the city. As the research goal shifted, our methodological approach became more focused on how we could gain insights into the lives of our mobile research subjects (Fincham et al., 2010). Figure 4.2 shows a group of drivers whom we met on our second day at stage on the outskirts of the city. We were riding with some drivers from the neighbouring stage, which got us an introduction. Once the novelty of two Muzungu (white people) from Manchester had worn off and we explained what we were doing, Mr Wize asked why he should talk to us. The idea that research may help put the case of the boda boda industry more effectively and give drivers a voice was instantly convincing.

We wanted to complement GPS tracks of boda bodas with video material collected using a GoPro mounted on the motorbike and a helmet camera to capture the driver's head movements (Spinney, 2011). In practice, the technology was far more useful as a way to engage our research participants. Our first day in the field involved hiring boda bodas to take us to their stages on the outskirts of Kampala as we specifically wanted to understand the role of boda bodas outside of the urban centre. A key challenge involved engaging with our drivers, who spoke little English and seemed uncomfortable in the company of Muzungu and the formal office setting. This changed when we got the GoPro camera out of our bag and asked if we could attach it to the boda boda's bike. Intrigued by the technology, our driver started helping attach it to the handlebars, and we soon had a small crowd of drivers around us (Figure 4.3). While the use of technology has been criticised by some in the mobilities field (e.g. Merriman, 2014), it played an essential role engaging our participants as co-producers of knowledge. This was the first time we had used the camera in the field, and we were both novices with the technology. This circumstance actually created a joint sense of achievement and amazement between

Figure 4.2 'Nobody listens to us.'

Source: Author's photo of Buwaate stage in front of a café, 07/12/2015.

Figure 4.3 Driver helping us attach a GoPro to his bike, with audience

Source: Author's photo.

the researchers and the boda drivers when we managed to connect it to our smart phone to view and control the camera feed. This seemed to be co-production of a different kind, not only setting aims and working together, but actually figuring out the best technical solutions in the field based upon the driver's experience of the motorbike's frame and the likely impacts of bumpy roads on the camera fixing. Vannini and Stewart (2016) have suggested that GoPros are able to disrupt the traditional tourist gaze by involving research participants more deeply in collecting a more personal form of data, and we found the use of GoPros created a similar effect among our drivers. The helmet camera also created interest and a degree of competition among drivers to be the ones who would wear it. When we arrived at various stages the technology operated as an entrée and a leveller, enabling us to speak to more drivers in a more open and equal fashion.

On a rainy Wednesday we once again made our way to the Tugende compound and spent an entire day organising our army of 26 surveyors (perhaps five times the number we planned for) who had been recruited by enthusiastic Tugende employees on our behalf. The result was 532 completed driver surveys from across the five districts of the city covering basic information about the driver, safety (covering their record, equipment and concerns), use of their boda boda by customers and journey type, and finally open-ended questions asking about the best and worst features of being a driver, and how they would improve and/ or regulate the industry. The survey was piloted on new drivers at a Tugende application day, which resulted in further refinement of the question phrasing. These data were complemented by 1,050 driver application forms supplied by Tugende, which although being far simpler asked some overlapping questions that enabled a degree of validation of the data.

The survey was originally intended to be a very small element of the research, but ended up taking up at least a third of our time in the field. We worried that the messiness of the research management process (Jones and Evans, 2011a) was deflecting us from our original goals as they had been stated on paper. As so much of Kampala is constructed through the noise, smell, dust and general perpetual chaos of the transport infrastructure, it could not fail to shape the way the research unfolded. We were certainly not following the gold standard of an 'ideology of objectivity' (Haraway, 1988: 584) whereby the researcher is positioned as an omnipotent expert in control of both passive research subjects and the research process, perhaps informed by years of positivist-inspired training which has taught us that 'impersonal, neutral detachment is an important criterion for good research' (England, 1994: 242). Despite much critique from feminist scholars (see McDowell, 1992; Rose, 1995; Sultana, 2007), rarely do we specifically consider the role of the researcher(s) as a messy one that has to be continuously (re)negotiated in the design, execution and delivery of research, but also one that could significantly enhance the research experience for all involved. Co-producing the research with Tugende, and subsequently the boda drivers themselves, imbued the research with a meaning and greater sense of mission than would otherwise have happened.

Using technology to capture moving subjects

The key element that we wanted to pursue in the project was the use of GPS and video to help capture our mobile subject matter. A rich body of work has developed around what Sheller and Urry (2006) identified 10 years ago as a mobilities paradigm, which recognised how many aspects of our lives have become more mobile through new ICT and travel technologies. A wide range of so-called mobile methods have subsequently been developed across the social sciences to try to capture both existing and new forms of mobility.

A considerable literature has developed around active transport, which includes walking and cycling, to understand how people experience places through these modes of movement, and what lessons this might hold for the design and planning of these places. Many of these studies have sought to bring qualitative methods together with GPS and Geographic Information System (GIS) technologies to better understand the relationship between peoples and their movements (Cope and Elwood, 2009; Jung and Elwood, 2010; Ricketts-Hein et al., 2008). The boda mobility project was in part inspired by a previous piece of research on 'Rescue Geography', which used walking interviews where the researcher accompanies the interviewee on a route around their local area, to understand resident attachments to place (Evans and Jones, 2011; Jones and Evans, 2011b). By mapping opinions and values, the project discovered that walking interviews produce richer data relating to place than traditional interviews. The method has also been applied to work with cyclists to identify how they experience cycling through specific urban areas and highlight potential challenges to encouraging this mode of transport. In the boda mobility project, we wanted to use qualitative GIS techniques to enable drivers to capture their daily experiences and subsequently represent them spatially through a map.

The project aimed to recruit boda boda drivers to carry GPS-enabled smart phones that would geospatially track their journeys and record them using a simple app. An added motivation for using this technique in the context of the more informal parts of Kampala is that it can identify routes that are not shown on maps (Hagen, 2011). Technically, the approach worked well. Kampala has a well-developed 4G+ network (we used MTN), and the assisted GPS in mobile phones uses the masts to give much faster GPS lock and improved accuracy to within roughly 7.5 meters in an urban environment like Kampala with few high-rise buildings (Evans and Jones, 2011b). Using smart phones played a huge role in both recruiting participants and engaging them with the project. Smart phones are universally desirable objects, and are ubiquitous in much of the Global South. There is no landline system, while Internet connectivity is still relatively low, so phones are used to fulfil a huge range of functions, including transferring money and paying bills. Perhaps even more importantly they are a key status symbol. Kirunda Brian, the first driver whom we recruited for this element of the research, was very enthusiastic about the idea that he could use a smart phone to contribute to the research. Twenty minutes and three miles around the Lugogo bypass later we were in a shop purchasing a phone with the right capabilities.

From this point Brian became an active researcher in the project, using the phone to take pictures he felt reflected his experiences as a boda boda driver. This kind of approach has been advocated by previous research on African mobilities (Robson et al., 2009), and as a co-researcher we asked Brian to try to capture images of motorcycles carrying various types of cargo, because we were interested in this specific role of boda bodas. Brian also focused on photographs of his preferred washing station, of various maintenance tasks being conducted by his mechanic, and of his stage and fellow stage members (Figure 4.4 a–c). These images generated considerable insights into the way boda bodas are embedded in the city. For example, washing the motorbikes is a major side industry, with numerous washing stations established throughout the city and staffed by an army of professional motorbike washers. Much of south and east Africa is characterised by murram roads, made of compacted dirt that during the dry season coats everything with a layer of red dust that can be abrasive and damaging and during the wet season creates a cloying quagmire. When you 'eat from your bike' there is a practical need to keep your bike clean and thus functional. Moreover, a clean motorbike is a real point of fare, enticing pride for boda boda drivers. Drivers would take an opportunity to clean accumulated dirt off their bikes throughout the day, often paying women at the side of the road for water to do so, but the 'main clean' was outsourced to these washing stations. The washing stations themselves offered a range of standards at corresponding costs. Some of these stations were located at

Figure 4.4 Photographs taken of (a) drivers of a washing station, (b) an oil change and (c) a milk delivery boda

Source: Photos reproduced with permission from Kirunda Brian.

Figure 4.4 Continued

roadsides next to a drain, waterway or swamp, while others were higher end, with proper forecourts and restaurant and bar areas to provide refreshments while drivers waited. Brain's photographs identified one unsuspected swamp as a key research site.

The main purpose of the smart phones was to capture time stamped and geo-tagged data about where the boda boda had been and when. Our original aim was to analyse this data in a fairly rigorous fashion using GIS to understand how boda bodas permeate the city, the routes that they are able to navigate and how this contributes to the daily needs of Kampalans. In practice, the primary advantage of mapping the daily routes was to act as an *aide memoire* for the drivers to help them remember the various trips and fares they had made that day. The fact that drivers would simply not remember their daily movements was an issue that was flagged to us on our second day in Kampala, when we presented our plans to the staff of Tugende, and this turned out to be entirely correct. Without being able to show the drivers their GPS track on a Google map of Kampala we would have been unable to collect any reliable or complete data concerning the average day of a driver. We would meet our driver at the end of their shift at the Tugende offices, and then download the GPS track from the phone onto a laptop. It was a fairly quick process to then open the track in Google Maps and get the driver to talk us through their day (Figure 4.5). We used a digital recorder to record this conversation, and created pins on the Google map to provide basic information about the nature of the trip and why the passenger was using the boda boda. Figure 4.6 gives an example of one of these tracks.

Using the map as the basis for debriefing drivers was not only essential in capturing the daily experiences and uses of a boda boda, but also revealed real depth about how they facilitate Kampalan life. One of Brian's regular jobs involved collecting cakes from a bakery in another district and taking them to the local market for sale. Another regular customer traded in second-hand TVs, and would invite Brian in to give opinions on various possible purchases on these trips, while

Figure 4.5 Kirunda Brian and Jennifer discussing the day's GPS tracks

Source: Author's photo.

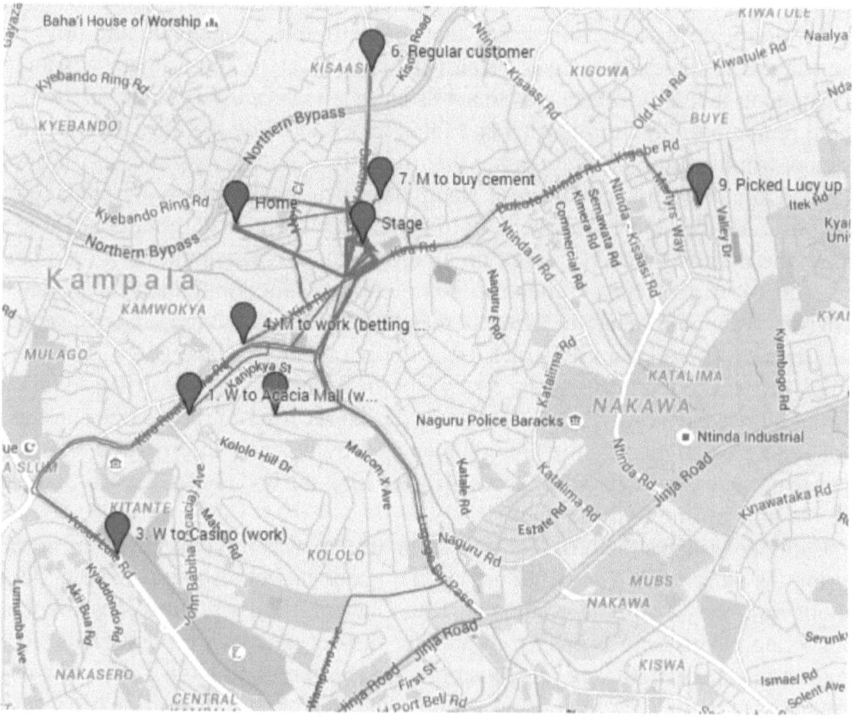

Figure 4.6 Kirunda Brian's GPS tracks for a typical shift

another would phone Brian to ask him to go and fetch breakfast from a rotating list of eateries. He would often collect cash from customers, go to a seller to purchase construction materials and then deliver them to a particular site. On another occasion, he collected money for a 'house girl' (a domestic servant usually from the country) before then picking her up from the supplier and dropping her at her new employer. Beyond these insights into the surprising and diverse ways that boda bodas facilitate Kampalan life, the overall patterns of activity were similar to those identified by the driver and passenger surveys we conducted and the broader GPS analysis of 400 boda boda tracks. About half of boda boda customers are regulars, with most of these using the boda bodas to get to or from work. About a quarter of all trips relate to deliveries of some form.

Discussing the tracks with the drivers also gave insights into their experiences, including the myriad decisions that they make every day concerning where to go, what routes to take at different times of the day and in different weather conditions, how long to wait and how to balance work with various personal errands. A major benefit of qualitative GIS is that it allows data to be represented in highly visual ways. Annotated GPS tracks contain text from the drivers' journey audio notes, interviews, photos and video material and can be made available via Google Maps in accessible formats that allow users to customise and interact with the data in simple yet powerful ways.

Conclusions

In his account of inner-city life in Johannesburg, AbdouMaliq Simone talks about 'people as infrastructure', which he intends to be suggestive of the 'collaboration among residents seemingly marginalized from and immiserated by urban life' that enables them to provide services and trade with one another (2004a: 407). Manifesting what Simone (2011) calls the distinctive mobility of the African city, where movement is essential to daily survival, boda bodas support the thickening fields of social relations that people depend on (ibid., 2004b). Boda mobility is generated from the intersection of bodies, technologies and landscapes and cannot be reduced to any one of these components. Although there is a current profusion of work engaging with the informality of African cities (Pieterse and Simone, 2013), new thinking is required on urban mobility (Pirie, 2014). We believe that qualitative GPS techniques offer an important approach to informal mobility combining methods from transport studies and mobilities research (Shaw and Hesse, 2010) by capturing the ways in which people, machines and urban landscapes co-constitute one another through their circulation.

A recent study project predicts that by 2100 five of the largest 10 cities in the world will be located in Africa, with Kampala projected to increase twentyfold to 40 million inhabitants (Hoornweg and Pope, 2013). Even allowing for considerable margins of error, this kind of explosive urban growth will be accompanied by massive increases in the prevalence and importance of informal transport in enabling mobility. Although the idea that informal transport is more adaptable and potentially sustainable is not new, conceptualising how the forms of mobility that it enables socially and materially reproduce urban landscapes is innovative and important. Boda bodas are part of the 'rogue' urbanism of African cities, 'unruly, unpredictable, surprising, confounding, and yet pregnant with possibility' (Pieterse and Simone, 2013: 12). The challenge to researchers, activists and policymakers lies in how to engage with such spaces and infrastructures that are characterised by informality and provisionality. While cognisant of the informal political structures through which boda bodas are organised, focusing on their materiality opens up new ways to think about mobility, and by extension the city itself (Dovey, 2011). This is important for sustainable mobility and by extension the future of cities and those who live in them.

Our experiences highlight three methodological insights that we think are of wider relevance to the study of informal mobilities. First, far from being a gimmick or bolted-on 'innovative element', technology played essential roles in our study, as an ice-breaker to engage participants, an *aide memoire* and a basis for co-production. It simply would not have been possible to gain meaningful insights into the mobile lives of boda boda drivers without these technologies. Second, it is hard to access the lives of those involved in providing informal transport using formal approaches. GPS tracks and formal surveys can provide important descriptive information and identify patterns of movement, but more informal and participatory techniques are required to show how these are generated through the mobile lives of drivers and passengers. Part of this involves being a reflexive researcher in order to better understand people's lived experiences, and be open to new research opportunities presented by a co-production approach. Finally, while motivated primarily by calls within African urbanism for more empirically rich and materially grounded analyses (Swilling et al., 2013), these kinds of studies of informal mobility are equally rare in the Global North. Although there is a need to expand the mobilities approach into the Global South, much can be learnt from both North and South.

References

Behrens, R., McCormick, D., & Mfinanga, D. (2016). *Paratransit in African Cities - Operations, Regulations and Reform*. Oxon: Routledge.

Cervero, R. and Golub, A. (2007) Informal transport: A global perspective. *Transport Policy* 14 (6), 445–457.

Cope, M. and Elwood, S. (2009) *Qualitative GIS: A Mixed Methods Approach*. London: Sage.

Cresswell, T. (2011) Mobilities I: Catching up. *Progress in Human Geography* 35 (4), 550–558.

Diaz Olvera, L., Plat, D., Pochet, P. and Maidadi, S. (2012). Motorbike taxis in the transport crisis of West and Central African Cities. July 13, *EchoGeo* 20.

DigitalMatatus (2014). *DigitalMatatus*. Available at: www.digitalmatatus.com/ (accessed 17 February 2014).

Dovey, K. (2011) Uprooting critical urbanism. *City* 15 (3–4), 347–354.

England, K. (1994) Getting personal: Reflexivity, positionality, and feminist research. *Professional Geographer* 46 (1), 80–89.

Evans, J. and Jones, P. (2011) The walking interview: Methodology, mobility and place. *Applied Geography* 31, 849–858.

Fincham, B., McGuinness, M. and Murray, L. (eds.), 2009. Mobile methodologies. London: Springer.

Goodfellow, T. and Titeca, K. (2012) Presidential intervention and the changing 'politics of survival' in Kampala's informal economy. *Cities* 29 (4), 264–270.

Hagen, E. (2011) Mapping change: Community information empowerment in Kibera. *Innovations: Technology, Governance, Globalisation* 6 (1), 69–94.

Haraway, D. (1988) Situated knowledges: The science question in feminism and the privilege of partial perspective. *Feminist Studies* 14 (3), 575–599.

Hoornweg, D. and Pope, K. (2013) *Socio Economic Pathways and Regional Distribution of the World's 101 Largest Cities*. No.4 Global Cities Institute Working Paper, University of Ontario Institute of Technology, Oshawa, Ontario.

Howe, J. and Davis, A. (2002) Boda boda – Uganda's rural and Urban low-capacity transport services. In: Godard, X. and Fatonzoun, I. (eds.), *Urban Mobility for All*. Lisse, 235–240. The Netherlands: Swets and Zeitlinger.

Jones, P. and Evans, J. (2011a). The spatial transcript: Analysing mobilities through qualitative GIS. *Area* 44 (1), 92–99.

Jones, P. and Evans, J. (2011b) Creativity and project management: A comic. *ACME: An International e-Journal for Critical Geographies* 10 (3), 585–632.

Jung, J. and Elwood, S. (2010) Extending the qualitative capabilities of GIS: Computer-aided qualitative GIS. *Transactions in GIS* 14 (1), 63–87.

Kampala Capital City Authority (2012) *Updating Kampala Structure Plan and Upgrading the Kampala GIS Unit*. Available at: www.kcca.go.ug/uploads/KPDP%20Draft%20 Final%20Report.pdf (accessed 6 January 2017).

Kampala Capital City Authority (2013). *Promoting Non-Motorised Transport: Case Study of the NMT Pilot Corridor*. Paper presented at the UNEP Share the Road, Nairobi, Kenya, October 29, 2010.

Kamuhanda, R. and Schmidt, O. (2009) Matatu: A case study of the core segment of the public transport market of Kampala, Uganda. *Transport Reviews* 29 (1), 129–142.

Kisaalita, W. and Sentongo-Kibalama, J. (2007) Delivery of urban transport in developing countries: The case for the motorcycle taxi service (boda-boda) operators of Kampala. *Development Southern Africa* 24 (2), 345–357.

Luthra, A. (2006) Para transit system in medium sized cities. *LTPI Journal* 3 (2), 55–61.

Magoola, J. (2013) Walk this way: Pedestrian road safety must be stepped up worldwide. *The Guardian*, October 23. Available at: www.theguardian.com/global-development/poverty-matters/2013/oct/24/walk-pedestrian-road-safety?CMP=twt_gu (accessed 28 October 2013).

McDowell, L. (1992) Doing gender: Feminism, feminists and research methods in human geography. *Transactions of the Institute of British Geographers* 17, 399–416.

Merriman, P. (2014) Rethinking mobile methods. *Mobilities* 9 (2), 167–187.

Pieterse, E. and Simone, A. (2013) *Rogue Urbanism: Emergent African Cities*. Johannesburg: Jacana Publishers.

Pirie, G. (2014) Transport pressures in urban Africa: Practices, policies, perspectives. In Pieterse, E. and Parnell, S. (eds.), *Africa's Urban Revolution*, 133–147. London: Zed Press.

Red Pepper (2013). *KCCA Finalises Plans to Throw 145,000 Boda-Boda Cyclists Out of Kampala*, 18 May. Available at: www.redpepper.co.ug/kcca-finalises-plans-to-throw-145000-boda-boda-cyclists-out-of-kampala/ (accessed 8 October 2013).

Ricketts-Hein, J., Evans, J. and Jones, P. (2008) Mobile methodologies: Theory, technology and practice. *Geography Compass* 2 (5), 1266–1285.

Robson, E., Porter, G., Hampshire, K. and Bourdillon, M., 2009. 'Doing it right?': working with young researchers in Malawi to investigate children, transport and mobility. *Children's geographies*, 7 (4), 467–480.

Rose, G. (1995) Distance, surface, elsewhere: a feminist critique of the space of phallocentric self/knowledge. *Environment and Planning D: Society and Space* 13, 761–781.

Shaw, J. and Hesse, M. (2010) Transport, geography and the 'new' mobilities. *Transactions of the Institute of British Geographers* 35 (3), 305–312.

Sheller, M. and Urry, J. (2006) The new mobilities paradigm. *Environment and Planning A* 38, 207–226.

Sietchiping, R., Permezel, M. and Ngomsi, C. (2012) Transport and mobility in sub-Saharan African cities. *Cities* 29, 183–189.

Simone, A. (2004a). People as infrastructure: Intersecting fragments in Johannesburg. *Public Culture* 16 (3), 407–429.

Simone, A. (2004b) *For the City Yet to Come: Changing African Life in Four Cities*. Durham, NC: Duke University Press.

Simone, A. (2011) The urbanity of movement. *Journal of Planning Education and Research* 31 (4), 379–391.

Spinney, J. (2011). A chance to catch a breath: Using mobile video ethnography in cycling research. *Mobilities* 6, 161–182.

Sultana, F. (2007) Reflexivity, positionality and participatory ethics: Negotiating fieldwork Dilemmas in international research. *Acme* 6 (3), 374–385.

Swilling, M., Robinson, B., Marvin, S. and Hodson, M. (2013) *City-Level Decoupling: Urban Resource Flows and the Politics of Infrastructure Transitions*. Nairobi: UNEP.

UN Habitat (2010) *Sustainable Mobility in African Cities*. Nairobi: UNEP.

Vannini, P. and Stewart, L. (2016) The GoPro Gaze. *Cultural Geographies* 24 (1), 149–155.

Williams, S., White, A., Waiganjo, P., Orwa, D. and Klopp, J. (2015) The digital Matatu project: Using cell phones to create an open source data for Nairobi's semi-formal bus system. *Journal of Transport Geography* 49, 39–51.

5 The paratransit puzzle

Mapping and master planning for transportation in Maputo and Nairobi

Jacqueline M Klopp and Clemence M Cavoli

Most citizens in African cities rely on semi-formal bus systems called paratransit. Yet these mobility systems tend to be invisible or marginal within formal urban planning processes: a "paratransit puzzle". Through comparative analysis of recent master plans for Maputo and Nairobi, we concretely demonstrate this gap in official planning. We also show preliminary evidence from Maputo and Nairobi to suggest that collaborative paratransit mapping can support more inclusive transportation planning conversations. Overall, given the importance of these mobility systems in African cities, we argue that paratransit needs to be a more central concern within an inclusive planning process.

Introduction

This chapter looks at what we call the "paratransit puzzle" in African transportation planning. Most people walk a great deal to access work and services in African cities. For longer distances many citizens rely heavily on "paratransit" – minibuses with flexible stops, routes and schedules (Behrens et al., 2016). Yet, as we show for the cities of Maputo and Nairobi, these mobility systems tend to be invisible or marginal within formal urban planning processes. Given how important and ubiquitous diverse forms of these mobility systems are in most African cities, this marginality of paratransit in planning is a puzzle.

Growing numbers of wealthier citizens are turning away from these urban mobility systems to purchase and use private cars for their daily commute and activity, as well as for social status. Facilities for taking the bus, walking, cycling or "non-motorized transport" (NMT) remain very poor and often unsafe (Khayesi et al., 2010, Mitullah et. al, 2017). NMT networks tend to be badly connected to bus stops and terminals where people gather to move larger distances across cities, helping to make African cities "disconnected and costly" (Lall. et al. 2017). Poor walkability, public transport and the rising use of personal cars contribute to the complex set of problems these cities are facing, including a rise in congestion; air pollution; noise; road crashes, in particular pedestrian fatalities; and social and spatial inequality among other problems (Pieterse, 2010; World Health Organisation, 2013).

By looking at the "paratransit puzzle" we ask: What kind of transportation planning will address these pressing problems, and how does it compare to the kind of planning experienced in African cities today? Can a transportation planning that fully embraces indigenous mobility systems and the practices of paratransit systems be part of the solution? And if so, can collaborative mapping projects help put paratransit on the map and more squarely in planning conversations for these cities?

Until recently, much academic work on transportation planning in sub-Saharan Africa has focused narrowly on engineering issues in relation to roads (Porter, 2007). A small, but rich, growing literature on African paratransit is emerging (Behrens et al., 2016; Khayesi et al., 2015; Hart, 2016; Klopp et al., 2015; Mutongi, 2006; Mutongi, 2017, Rizzo, 2017, Schalecamp and Behrens, 2013; Williams et al., 2015; Woolf and Joubert, 2013) along with work on non-motorized transport (Khayesi et al., 2010; Howe, 1995; Mitullah et al., 2017). However, a bias towards road construction and motorized private mobility seems to exist, rather than a more people-focused concern with access and equity as well as public health and sustainability, when it comes to actual planning and investments on the ground (Mitric, 2013; Klopp, 2012; Sietchping et al., 2012).

The extent to which a shift in planning practices is occurring to address the increasingly well-documented skewed orientation of transportation planning in African cities is an open and important question. In this chapter, we cast a critical, comparative eye on two recent master planning efforts in Maputo and Nairobi and how they address paratransit. After a brief look at Maputo and Nairobi, we analyze their master plans, with a critical focus on how paratransit is treated within these plans. With assistance from the Japanese International Cooperation Agency (JICA), both cities completed master plans in 2014, offering an important comparative opportunity to explore how planning processes operate in practice and the extent to which the plans produced in this way addresses the transportation and accessibility needs of the majority of people living in these cities. We also discuss how the bottom-up process of mapping paratransit in these cities via Global Positioning System (GPS) tracking and elaborate collaboration with diverse actors through the Mapa Dos Chapas in Maputo and the Digital Matatus works. The idea of these projects is to use the mapping to re-center planning around the existing paratransit systems in these cities. Finally, we ask whether these projects are having the intended impact and the way forward for an improved planning process for African cities.

Maputo and Nairobi

Nairobi and Maputo have different and complex histories. However, both cities emerged and grew as transportation hubs within a context of conquest and colonialism. Nairobi began as an outpost and then a small town linked to the headquarters of the Uganda Railways. Maputo started as a small garrison port town of Portuguese soldiers and traders strategically located at the meeting place of what is now called the Maputo River and Maputo Bay. Like Nairobi, Maputo's growth as a city was also linked to a railway project: the Transvaal Railway was built to connect strategic regions in Southern Africa to the port (Vanin, 2013).

Both cities are "dual" or "divided" cities marked by extensive poor neighbourhoods separated by distance, physical structure or security forces from exclusive areas and gated communities (Paasche et al., 2010; Olima, 1997). In both cities, areas that have been planned initially for the colonial elite (in Maputo the "concrete city" or cidade de cimento) tend to stand in contrast to the extensive, poorly organized "informal" settlements where the poor and lower-middle class tend to live (Vanin, 2013; King, 1976). Today, what was once racial and social segregation in the city is now primarily social segregation made worse by the growth of exclusive communities and "cities" (Olima, 1997; Morange et al., 2012).

This dualistic structure of the city shapes and reflects the transportation dynamics in both cities. Although some limited public transport exists (Nairobi commuter rail and the Maputo Bus Company buses operating on 60 routes), wealthier residents use automobiles and indigenous entrepreneurs run the minibus systems, which serve the neglected majority of people living in poor/lower middle–class regions of the city (Mutongi, 2006; Khayesi et al., 2015). Both cities reflect the more general trend of formal bus system decline in the face of underinvestment, political manoeuvring and competition from the minibus system (Rizzo, 2002). Maputo, of course, had the added impact of a long civil war (1977–1992) and although investment in public buses increased since the 2000s, formal buses still serve substantially fewer citizens than chapas in the capital city (JICA, 2014d).

Today, Nairobi is a bustling, very rapidly growing city of approximately 3.4 million people, and more than 7.5 million in the wider metropolitan area (JICA, 2014a). Maputo has a population of approximately 1.2 million people, whilst Greater Maputo (including neighbouring Matola, Boane and Marracuene suburbs) has more than 2.2 million inhabitants and is expected to nearly double by 2035 (JICA, 2014b). In both cities, large numbers of people live in slums and the poorest walk as the main mode of transport, while paratransit – matatus or chapas – provides the backbone of mass public transit in both cities. Motorization rates also continue to grow, which along with poor roads and poor traffic management, mean worsening road congestion, which is already considered severe in Nairobi and is rapidly becoming a problem in Maputo. Air pollution and road safety are already serious concerns in both cities (World Health Organization, 2013, 2016; Gascon et al., 2016; Kinney et al., 2011; Nizamo et al., 2006).

Rendering paratransit visible through mapping

Digital matatus project in Nairobi

In 2011, a consortium of universities (University of Nairobi, MIT, Columbia University) and a small design firm specializing in informality (Groupshot) began a project of using geo-location–enabled cellphones and GPS units to map the matatu system in Nairobi. The team included mappers from the University of Nairobi who were computer science students and regular matatu users. By talking to drivers and passengers at key terminals and stations and riding different routes multiple times with GPS-enabled cellphones, they were able to capture the

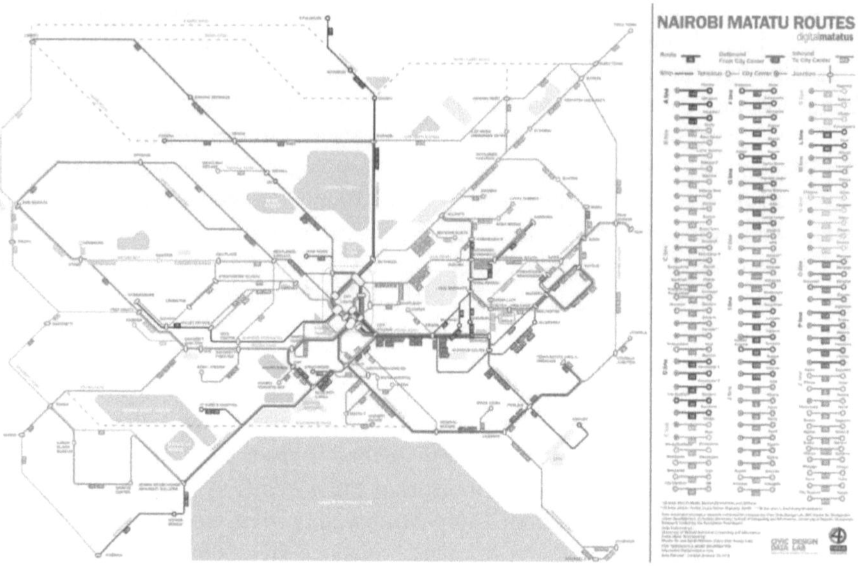

Figure 5.1 Nairobi Matatu Map

Source: Courtesy of Digital Matatus/Civic Data Design Lab, MIT.

route and also where regular stops are located (Williams et al., 2015; Klopp et al., 2015). These data enabled the creation of a public transit map (Figure 5.1) for the city by the Civic Data Design Lab at MIT.

This mapping project set out to engage a wide variety of actors from government, civil society and the technology and transportation sector throughout the process to get feedback and build understanding of the data including its standardized format (General Transit Feed Specification) and hence to build users as well as have policy and planning impact. This mapping was part of a specific pro-urban public transport advocacy agenda influenced by a growing civic hackers' movement to improve user interaction with public transport options by providing high-quality information through apps and maps (Townsend, 2013; McHugh, 2013).

Despite many efforts to engage with the government and also multi-lateral funders – the "official" planners for Nairobi – support for the mapping largely came from young tech-savvy technocrats in the government think tank, the Kenya Institute for Public Policy Research and Analysis (KIPPRA), and the Kenya Open Data Initiative. The key actors involved in transportation planning for Nairobi – the transportation department of the City County of Nairobi and the National Transport and Safety Authority and the Ministry of Transportation and Infrastructure– were considerably more hesitant to embrace the idea of open matatu data, most likely because of the complex informal politics around route allocation, which raises significant funds for the government in the form of licenses. The data and map raise many transparency questions and also highlight the abdication of planning by the government. In contrast, passengers and drivers were supportive of the mapping process, and interviews and focus group discussions with drivers revealed that there was considerable pride around the extent of mobility services they provide, represented by the many routes stretching across the city.

In January 2014, officials from the governor's office of Nairobi City County and the Ministry of Transport and Infrastructure (MOTI) along with the Matatu Owners Association came to a data and public map launch. The governor of the city would later make public pronouncements of support with language that appeared to give the city some credit for the Digital Matatus efforts. However, the government has in fact been slow to officially embrace the data. Multilateral actors/lenders such as the World Bank, the European Union and the African Development Bank and their consultants, are increasingly using the data and maps, for example, to study spatial mismatch and access to health facilities in Nairobi (Avner and Lall, 2016; Cira et al., 2016). However, these institutions do not generally recognize a role in or responsibility for supporting the creation of such data as part of support for transportation planning.

An exception is UN-Habitat, which after engaging in a number of conversations with Digital Matatus engaged the Institute of Transportation Development Policy (ITDP) to replicate the mapping process in Kampala. Later, ITDP would get a grant to map out the Kenyan city of Kisumu. UN-Habitat and ITDP also used the digital matatus data for their Bus Rapid Transit (BRT) service plan for Nairobi, saving time and effort; in this practical way, they experienced the value of high-quality open minibus data (UN-Habitat and ITDP, 2015). Still, the problem of negotiating with the matatu sector and building a strong process of engagement, which this kind of mapping can encourage, unfortunately, does not seem as much of a priority as the technical designs.

Mapa Dos Chapas (public transport map of Maputo)

Initiated in June 2014, the "Mapa Dos Chapas" project aims to give visibility to the chapas system by collecting minibus routes and bus stops data and translating them into a user-friendly map (latest version of the map is depicted in Figure 5.2). It also aims to generate conversations about urban mobility in Maputo and the role of the chapas in the transport network. The project team includes international consultants, academics (based at University College London) and Associação Rede Uthende (RUTH), a local Mozambican association that supports and promotes participatory governance and citizens' advocacy. The involvement of RUTH was particularly important to legitimize the project and initiate bottom-up processes.

Learning from the experiences of the Digital Matatus project, the mapping process was undertaken in close collaboration with the main chapas associations. The first step of the mapping process consisted of collecting GPS coordinates of minibus routes and commonly used informal minibus stops. The data were then compiled, verified and approved by the chapas representatives. This is important,

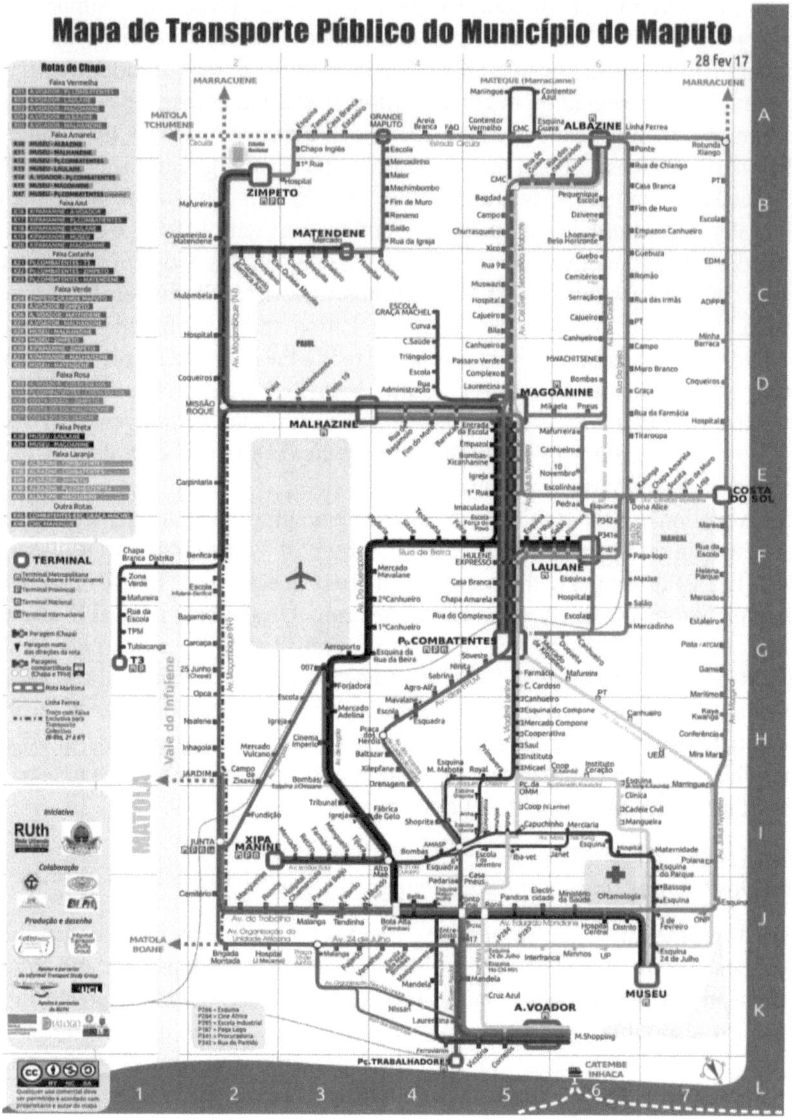

Figure 5.2 The public chapas map for Maputo

as it represents a step towards the formalization of a structured system. Indeed, the formal recognition of a set of routes and bus stops is likely to lead chapas owners and drivers to adhere to the routes and stops agreed to in the context of the mapping process. As in Nairobi, the chapas association representatives view the map as an opportunity to enhance the visibility of their service and to gain recognition. As a result, they have been actively supporting the Mapa Dos Chapas project.

From an early stage, a key element of the project was the engagement with local authorities in Maputo, in particular the Maputo City Council Transport Department. Unlike the Nairobi project, the Mapa Dos Chapas team sought formal approval of the mapping process and of the map from the public authorities in charge of transport in the capital city. After a lengthy process the Maputo City Council approved the map and embraced the mapping process as an opportunity to collect valuable data and have a better understanding – and potentially control – of the chapas system.

Since the establishment of a draft map, the chapas map has been used as a working document by both the local authority and the chapas associations, in particular when new chapas routes were licensed in March 2016. It is noticeable that since the beginning of 2016 transport authorities in Maputo have given increased importance to the chapas system. The city council supported the creation of a cooperative representing chapas owners, called COOTRACK1, and granted this association the right to buy 50 government buses and operate them on the newly established dedicated bus lane in Maputo. Public authorities increasingly appear to be seeing chapas operators as important stakeholders in Maputo's mobility scene/transport system. Initial results suggest that the mapping process may have contributed to this recent change.

Two cities, one master plan dynamic

At the same time that the Digital Matatus project got started in Nairobi in 2102, JICA signed cooperation agreements for master plans with the Kenyan and Mozambique governments and work commenced shortly afterwards. The overt aims of these plans were similar. Initial 2012 meeting notes in Nairobi spelled out the rationale: "In order to accelerate sound and sustainable development, an integrated urban master plan has to be prepared to improve infrastructure such as transport network, water supply, sewerage reticulation energy, etc". (JICA, 2014c). The focus was also on places where projects could be conducted to address urban development issues. In Maputo, the aim was described initially in a similar way albeit focused only on urban transportation: the plan is "a way to create a comprehensive urban transport master plan for Greater Maputo for 2035, and conduct a pre-feasibility study for the priority projects"[1] (JICA, 2014d). In the case of Maputo, the pre-feasibility study was undertaken with the aim of establishing a bus rapid transit (BRT) system.

In both Maputo and Nairobi, JICA supported the large engineering firm Nippon Koei Co. Ltd to lead the master planning processes along with a number of other Japanese engineering and consulting firms.[2] It is an explicit aim of JICA to support "the entry of private-sector companies in developing countries" (JICA, 2016).[3] At the same time, "JICA claims to actively promote participatory cooperation with the residents and collaboration with NGOs with a focus on the beneficiaries, including users and residents, in view of who will use the system and for what purpose".[4] Some level of consultation occurred in both planning exercises.

In Maputo, the plan vaguely refers to "stakeholder interviews and discussions", but discussions with stakeholders revealed that consultations with minibus representatives were limited. In Nairobi, the consultation processes were quite extensive. Technical committees were set up around the different thematic areas, including transportation and land use. These technical working groups included experts who continuously reviewed and discussed the plan text. Through workshops and technical working group meetings, stakeholders provided reviews and input and neighbourhood/zone-based stakeholder participation forums were also organized (Gachanja, Personal Communication 2016).[5] Kenya has requirements for participation in its urban planning laws and constitution and underwent a transition to devolved power where the county of Nairobi gained greater powers. This may help explain the more extensive consultative process in Nairobi.

Regardless of these differences in process and scope, in the end, both plans share some common features. First, the transport data in these reports were collected by standard engineering methods, including travel surveys that ask households about their trips and passengers at certain locations about where they are going and why. These data reaffirmed how important paratransit is for both cities. In Nairobi, the data suggested matatus represent 44% of the modal share and move about 3.5 million commuters a day (JICA, 2014a). In Maputo, the plan noted that approximately 60% of all non-walking trips in Greater Maputo were made by chapas, and only 17% by conventional bus (JICA, 2014d: 29).

Secondly, both reports, despite recognizing the critical importance of paratransit, contain limited data of basic routes and stops of the paratransit system. Interestingly, little local involvement occurred in the data collection process and analysis for the Nairobi master plan. This work was done by Japanese consultants, and the software to do the modeling was not shared with local institutions (Gachanja, Personal Communication 2016). The data were also not made easily accessible limiting opportunities to be able to reproduce or challenge the analysis and its assumptions. In addition, data on the matatu routes and stops were widely

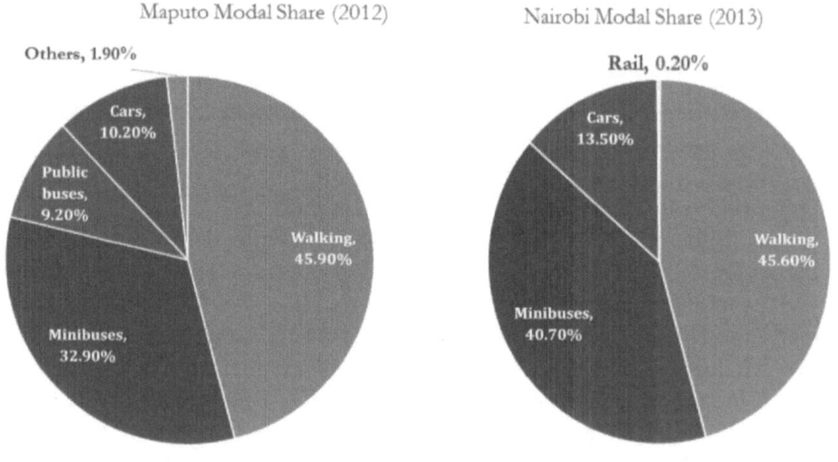

Figure 5.3 Maputo and Nairobi modal shares compared

available in Nairobi before the launch of the master plan thanks to the Digital Matatu project (Klopp et al., 2015; Williams et al., 2015), yet none of this critical transport data were used or cited in the Nairobi master plan.

In Maputo, the local transport plan produced by JICA contains limited data about minibuses, apart from the data about the modal share (collected via a household survey). In total, two paragraphs are dedicated to describing the minibus network and an incomplete map was used to illustrate some of the minibus routes. It is clear that the JICA plan contains insufficient data to fully integrate this mode of transportation into the wider transport system, in particular the BRT. Thanks to the recent Mapa Dos Chapas project we know there are about 45 chapas routes within Maputo and more than 140 in the metropolitan area (including Maputo, Matola, Marracuene and Boane). Attempts to collect data and conduct analysis on the dominant form of mass transportation in these cities just scraped the surface and did not generate sufficient data to properly understand, let alone integrate, those systems into future BRT plans.

Furthermore, the description of paratransit systems in both local transport plans focus primarily on the shortcomings of these systems without highlighting the benefits of those services to the city and their potential, and without suggesting any improvements in their work plans or any detailed integration with future transport systems. In Maputo, despite acknowledging that chapas rides represent 60% of all non-walking trips, the plan only highlights the drawbacks of the system, rather than the access to the city it generates. Chapas are primarily described as "old and poorly maintained" and the system is described as ineffective, with photographs highlighting the "long queue" at bus stops and the "dilapidated chapa vehicle". The summary of the chapas written in the transport plan reads: "Although some improvement has been found in enforcement of safety standards recently, various problems still remain, including: a lack of reliability, long waiting times, overcrowding, safety and security issues, inconvenient routing, poor accessibility, route cutting, forced interchanges, poor vehicle condition, low safety standards, and a lack of comfort" (JICA, 2014d).

Similarly, in the Nairobi plan, the matatu system is noted as a source of air pollution and congestion (JICA, 2014a). The plan also comments on the lack of perceived organization of the matatu system, noting that "outside the city centre, lay-bys for bus stops are found on the trunk roads, but along minor roads, matatus and buses often stop at roadsides or intersections to pick up passengers, which cause obstacles in the traffic flow of the roads". (JICA, 2014a: 35–36). However, to the credit of the consultants, they did ask passengers for their views on improvements, finding the following priorities: 1) improvement of bus stop facility/information, 2) improvement of accessibility and 3) improvement of regularity/punctuality. These are all elements that would improve the matatu system and the daily life of the 70% of Nairobi adults who use the system every day (Klopp and Rateng, 2015). The plan also includes improving the matatu system as an option. However, the authors provide few details, but include a project to relocate and revamp the main bus terminal to link it to transit-oriented development around an upgraded commuter rail station.

Although both of these reports do indeed discuss the importance of efficient public transportation and also in passing the need for NMT facilities – a step forward from past planning exercises – both focus on the larger projects including road building, bus rapid transit (Maputo) and monorail (Nairobi) instead of improving the systems that residents already use heavily. Inclusive mobility and access are not yet the main focus of this planning (Sclar et al., 2014; UN-Habitat, 2013)- even though in the Nairobi case, passengers focused on access as an issue. Overall, the view of paratransit is as an anachronism, and in Maputo explicit recommendations are made to "gradually replace chapas with large-size buses" (JICA, 2014d, 74) – or large-scale infrastructure such as rail and bus rapid transit. BRT in particular is a growing trend for many African cities because of what is perceived their relatively moderate cost in comparison to rail (Deng and Nelson, 2011). The Maputo plan does, however, recognize the potential need for minibuses in suburban areas to complement the future BRT system: "Mini-size buses can be used as feeders in densely populated suburban areas". (JICA, 2014d: 74). But the detailed BRT plan included in Maputo's local transport plan makes no explicit or detailed plan to integrate the future BRT into existing minibus networks or NMT facilities. The priority is clearly given to large public buses, and the plan does not clarify the extent to which chapas will remain part of the system.

The Maputo transport plan does recognize that phasing out the majority of the chapas is likely to create a political problem: "chapa operators, fearing for their livelihoods", will resist (JICA, 2014d: 68). JICA notes the need to offer adequate compensation to operators, including "the opportunity to own shares in newly established vehicle operating companies" and implementation of a job-creating program for chapa drivers (JICA, 2014d: 72). Also, the report notes that chapa drivers should be consulted and "involved in stakeholder meetings". Despite the plea for consultation with the chapas operators, comprehensive consultations did not seem to occur during the course of the planning exercise.

A somewhat similar but more inclusive vision is found in the Nairobi master plan. Three scenarios are constructed, ranging from road widening (car oriented) to a mass transit–oriented approach with an intermediate option in between. They advocate support for a massive mass transit push. This involves light rail, BRT and even a metro along the main corridors. Unlike the Maputo plan, the Nairobi master plan does not envision the replacement of paratransit. It foresees that "during the stage when a Mass Rapid Transit System (MRTS) is the major public transport in the city, existing matatus will operate as feeder service mode and cover the area which MRTS will not cover" (JICA, 2014b: 105). In the Nairobi plan, the power of the matatu industry to negotiate terms, stemming from its fundamental importance to the city and high levels of organization in associations, perhaps had an influence in the plea for consultations in the master plan much as in the Maputo case. The plan suggests matatus are less likely to be replaced, and they are integrated into the final public transit–oriented vision but only as "feeders".

Overall, both plans reveal how important paratransit systems are but the "puzzle" is that the planners did not go deeper to get more critical information on paratransit to inform the plans, even when more data could have been collected or was available. Although both plans mention the possibility that minibuses could operate as feeders, no specific or detailed planning is included in the BRT plans suggesting that in both cities the BRTs will be established without a detailed plan to include minibuses even as "feeders". This is more surprising because both plans recognize the political risks of failing to properly engage and include paratransit operators in upgrading public transport in these cities. Unless these systems are properly understood, engaged and integrated, it is hard to see how equitable, sustainable and inclusive improvements in public transport are to occur. Finally, even when planners conducted passenger surveys in Nairobi, and people expressed their desires for specific improvements in paratransit, the plan did not address these concerns.

In fairness, one barrier to suggesting more extensive projects to improve the paratransit system is the high level of informality and confusion about the legal status of routes and stops. Further, informality is politically negotiated and often involves an opaque set of irregular arrangements that generate benefits for some actors in the system as well as losses for others (Cervero, 2000; Klopp and Mitullah, 2016). For example, very poor conditions for workers emerge out of these arrangements (ILO, 2012; Ference, 2016). In the end, these poor working conditions, including long hours and low wages, allow these systems to operate without subsidies and in fact, make money for many actors, including those, like the police in Nairobi, who extract from the system through bribes (Klopp and Mitullah, 2016; Rasmussen, 2012). In addition, the complex and often chaotic-looking appearance of these paratransit systems makes it easier to simply assume they will be replaced and hence can be ignored within planning. Rather than discovering through engagement the ways in which these minibus mobility systems actually function and perform, it is more convenient to skirt around addressing paratransit in any detail.

Conclusions: towards an alternative planning paradigm

"Master" planning in African cities began within the colonial experience. Ever since it has been profoundly critiqued as a form of power (Njoh, 2009; Myers, 2011; Jacobs, 1961). Nevertheless, top-down "master" planning persists in modified forms up to present in African cities (Todes et al., 2010). Engineers, economists, foreign funders and consultants often dominate transportation planning, sometimes appearing like a "tyranny of experts" (Easterly, 2013). These transport planners tend to rely on particular methodologies and metrics that often marginalize the mobility and access concerns of the poorest in society who rely on NMT and paratransit for survival. They also marginalize minibus mobility systems that, despite their flaws, work to keep African cities functioning.

Confronting local political forces and diverse resistance from below, this form of master planning usually does not often succeed in realizing all the imagined projects and transformations in the plan. However, in some cases, the plans, which are increasingly prioritizing public transport and recognizing NMT, are actually better than what in the end gets implemented. Sadly, some of the public transport projects appear to serve as cover for ongoing road building as usual. Much of the road widening and building is now being promoted as part of this kind of larger urban transport vision, although a consistent pattern on the ground in Nairobi – with some exceptions – appears to involve ignoring this vision in the actual road designs, deprioritizing public transport and NMT and ignoring paratransit altogether. For example, the World Bank recently negotiated an elevated highway project in Nairobi, justifying the project by proposing a BRT to run on that highway. It is very likely that if this project goes forward, the BRT part might fail but a damaging highway expansion will occur.

Current top-down transportation planning succeeds in the ways best described by Ferguson's "Anti-Politics machine" (1994); master planning processes "work" for those within the planning enterprise itself and this in turn shapes the planning imaginary. In the Maputo and Nairobi cases, the involvement of big global consulting firms that specialize in large infrastructure projects appears to create a bias towards such projects. Large projects create the "economic and also a political sublime" – jobs and business as well as political visibility for those within the planning networks (Flyvbjerg, 2014). Government actors also find numerous ways to benefit and embed these projects in patronage dynamics (Klopp, 2012). In this way, master plans led by foreign consultants in collaboration with government agencies and their preferred consultancy firms produce the next set of contracts and work for themselves and in some ways, this is an explicit aim of JICA and many other "development partners" as well.

We see this process reflected in the kinds of data collected and the approach to analysis, where significant exclusions emerge in relation to paratransit. As we have stated previously, the planners collected very limited data about chapas and matatus – but the data that was collected pointed to how critical the services are to residents. In the Nairobi case, some specific desired improvements were expressed by commuters but then not wholly addressed in the plan. Overall, the positive aspects of the paratransit sector –its flexibility, its adaptability, the access it generates for the majority, its immense local knowledge of transport demand and traffic and even its role in providing local freight services – are not considered. The failure to open transport and land use data and make it more widely accessible also meant that technical arguments were hard to question even by local experts. Although NMT was also recognized as important, no data beyond what was gathered in the traditional household surveys was collected on this mode of the very poorest, and no independent projects to improve NMT facilities were included in the plans.

The large gap between existing transportation planning and investment and the mobility and access needs of the majority of African residents has wide-ranging ramifications. Failure to create high-quality, affordable transportation options and walkability to jobs and services in these cities helps to reproduce global patterns of automobile-dependent development (UN-Habitat, 2013). We know that this form of development will exacerbate current public health, equity and sustainability problems in African cities. We also know that improving rail and building

BRT while laudable, alone will never solve the access problems in the city. Part of the solution, as citizens often point out when consulted, has to be building on and upgrading paratransit and NMT. This must be a core part of designing a transport and mobility network focused on accessibility, sustainability and liveability. This in turn demands replacing the current top-down, highly dependent and often opaque transportation planning with a more inclusive planning. This more inclusive planning needs to be energized by bottom-up, locally rooted and led engagement processes where citizens, current transportation operators and local experts are not just consulted but are also actually heard.

Acknowledgements

We would like to give special thanks to Joaquin Romero de Tejada who coordinates the Maputo project locally, undertook the fieldwork in Maputo and provided data and valuable comments on this paper. We would also like to thank the Kestrelman Trust; their support has been key to the Maputo project. Thanks also go to Digital matatu team members Sarah Williams, Peter Waiganjo, Dan Orwa and Adam White as well as mappers at the University of Nairobi and James Gachanja for his valuable input to this work. Thanks also to the Rockefeller Foundation which made the Digital Matatus project possible and the Volvo Research and Education Foundations that supported the writing of this chapter. Last but not least, thanks to the many people in the paratransit and planning sectors in Maputo and Nairobi who took their time to talk to us. The ideas and thoughts expressed in this chapter are, of course, our own.

Notes

1 In Maputo, the local transport plan established by JICA is divided into two parts. The first part describes the city's profile and estimates future traffic demand. The main recommendation in the first part of the plan is to further develop the road network and to improve traffic management. The second part is a "Pre-feasibility study on the prioritized BRT project".

2 Other firms include the engineering firm PADECO Co. Ltd (Maputo), the International Development Center of Japan Inc. (the consulting arm of the non-profit research center by the same name) (Nairobi) and Eight Japanese Engineering Consultants Inc. (Nairobi).

3 Less directly connected to the planning process but no doubt relevant is the fact that Japanese vehicle manufacturers, especially Toyota and Nissan, are some of the main suppliers of vehicles – both used and new – in Mozambique and Kenya. Both countries are seen to be emerging and growing markets for private cars as well as other kinds of motor vehicles.

4 www.jica.go.jp/english/our_work/thematic_issues/transportation/activity.html

5 This same Kenyan expert noted, "Perhaps the key challenge which can be improved on is the technical aspects of data collection, analysis (modeling using JICA strada) and interpretation. The local team needs to be involved more effectively and capacity provided for open access to the Transport Modeling methods and software used. The software, hardware and derived model should be provided to NCC staff and stakeholders for continuous research, updating and revision of plan proposals (technology transfer)."

References

Avner, P. and Lall, S.V. (2016). *Matchmaking in Nairobi: The Role of Land Use*. Policy Working Paper. Washington, DC: World Bank.

Behrens, R., McCormick, D. and Mfinanga, D. (2016) *Paratransit in African Cities: Operations, Regulation and Reform*. New York, NY: Routledge.

Cervero, R. (2000) *Informal Transit in the Developing World*. Nairobi: United Nations Commission on Settlements.

Cira, D.A., Kamunyori, S.W. and Babijes, R.M. (2016) *Kenya Urbanization Review*. Washington, DC: World Bank.

Deng, T. and Nelson, J.D. (2011) Recent developments in bus rapid transit: A review of the literature. *Transport Reviews* 31 (1), 69–96.

Easterly, W. (2013) *The Tyranny of Experts: Economists, Dictators, and the Forgotten Rights of the Poor*. New York, NY: Perseus Books.

Ference, M. (2016) 'Together we can': Redefining work in Nairobi's Urban transportation sector. *Anthropology of Work Review* 37 (2), 101–112.

Ferguson, J. (1994) *The Anti-Politics Machine: Development, Depoliticization, and Bureaucratic Power in Lesotho*. Minneapolis: University of Minnesota Press.

Flyvbjerg, B. (2014) What you should know about megaprojects and why: An overview. *Project Management Journal* 45 (2), 6–19.Gachanja J. (2016) Senior Infrastructure Specialist, Kenya Institute of Public Policy and Research Analysis, Nairobi Personal Communication with authors.

Gascon, M., Rojas-Rueda, D., Torrico, S., Torrico, F., Manaca, M., Plasència, A. and Nieuwenhuijsen M. (2016) Urban policies and health in developing countries: The case of Maputo (Mozambique) and Cochabamba (Bolivia). *Public Health Open Journal* 1(2), 24–31.

Hagans, C. (2011). *Livelihoods, Location and Public Transport: Opportunities for Poverty Reduction and Risks of Splintering Urbanism in Nairobi's Spatial Planning*. Available at: www.bartlett.ucl.ac.uk/dpu/metrocables/dissemination/Hagans-2011.pdf.

Hart, J. (2016) *Ghana on the Go: African Mobility in the Age of Motor Transportation*. Bloomington and Indianapolis: Indiana University Press.

Howe, J. (1995) "Enhancing Nonmotorized Transportation Use in Africa-Changing the Policy Climate" Transportation Research Record. no 1487: 22-26

ILO, 2012. Baseline Survey of the Matatu Industry in Kenya. Geneva: ILO.

Jacobs, J. (1961) *The Death and Life of Great American Cities*. New York, NY: Vintage.

Jenkins, P. (2000). Urban management, urban poverty and urban governance: Planning and land management in Maputo. *Environment and Urbanization* 12 (1), 137–152.

Jenkins, P. and Wilkinson, P. (2002). Assessing the growing impact of the global economy on Urban development in Southern African cities: Case studies in Maputo and Cape Town. *Cities* 19 (1), 33–47.

JICA. (2014a). *The Project on Integrated Urban Development Master Plan for the City of Nairobi in the Republic of Kenya. Final Report. Part I. Current Conditions*. Available at: www.jica.go.jp/english/news/field/2015/c8h0vm0000966zqy-att/c8h0vm0000966zvl.pdf (accessed 10 August 2016).

JICA. (2014b). *The Project on Integrated Urban Development Master Plan for the City of Nairobi in the Republic of Kenya. Final Report. Part II The Master Plan*. Available at: www.jica.go.jp/english/news/field/2015/c8h0vm0000966zqy-att/c8h0vm0000966zvu.pdf (accessed 10 August 2016).

JICA. (2014c). *The Project on Integrated Urban Development Master Plan for the City of Nairobi in the Republic of Kenya. Final Report. Appendix*. Available at www.jica.go.jp/english/news/field/2015/c8h0vm0000966zqy-att/c8h0vm0000966zvx.pdf (accessed 10 August 2016).

JICA. (2014d) *Comprehensive Urban Transport Master Plan for Maputo. Final Report. Volume I.* Available at: http://open_jicareport.jica.go.jp/pdf/12152617_01.pdf (accessed 10 August 2016).

JICA. (2014e) *Comprehensive Urban Transport Master Plan for Maputo. Final Report. Volume II.* Available at: http://open_jicareport.jica.go.jp/pdf/12152625_01.pdf (accessed 10 August 2016).

JICA. (2016). *JICA Activities, Transportation* Available at: www.jica.go.jp/english/our_work/thematic_issues/transportation/activity.html (accessed August 2016).

Khayesi, M., Monheim, H. and Nebe, J.M. (2010) Negotiating 'streets for all' in urban transport planning: The case for pedestrians, cyclists and street vendors in Nairobi, Kenya. *Antipode* 42 (1), 103–126.

Khayesi, M., Muyia Nafukho, F. and Kemuma, J. (2015) *Informal Transport in Practice: Matatu Enterpreneurship.* Surrey: Ashgate Publishing.

King, A. (1976) *Colonial Urban Development.* New York, NY: Routledge.

Kinney, P.L., Gatari, M., Volavka-Close, N., Ngo, N., Ndiba, P.K., Law, A., Gachanja, A., Mwaniki, S., Chillurd, S.N. and Scalar, E. (2011) Traffic impacts on PM2.5 air quality in Nairobi, Kenya. *Environmental Science and Policy* 14, 369–378.

Klopp, J.M. (2012) Towards a political economy of transportation policy and practice in Nairobi. *Urban Forum* 23 (1), 1–21.

Klopp, J.M. and Mitullah, W. (2016) Politics, policy and paratransit: A view from Nairobi. In: Behrens, R., McCormick, D. and Mfinanga, D. (eds.), *Paratransit for African Cities.* London: Routledge.

Klopp, J.M. and Rateng, V. (2015) *Transport Sector Public Opinion Survey* Available at: www.slideshare.net/ipsoske/ke-ipsos-columbiauniversityreportjune2015pafinalversion (accessed 20 March 2017).

Klopp, J.M., Williams, S., Waiganjo, P., Dan Orwa, D. and White, A. (2015) Leveraging cellphones for wayfinding and journey planning in semi-formal bus systems: Lessons from digital matatus in Nairobi. *Planning Support Systems and Smart Cities* Springer, 227–241.

Lall, S.V., Henderson, J.V. and Venables, A.J. (2017) *Africa's Cities: Opening Doors to the World.* Washington, DC: World Bank.

McHugh, B. (2013). Pioneering open data standards: The GTFS Story. In: Goldstein, B. (ed.), *Beyond Transparency.* Available at: http://beyondtransparency.org/chapters/part-2/pioneering-open-data-standards-the-gtfs-story/ (accessed 20 March 2017).

Mitric, S. (2013) Urban transport lending by the World Bank: The last decade. *Research in Transportation Economics* 40, 19–33.

Mitullah, W., Vanderschuren, M. and Melecki, K. (eds.). (2017) *Non-Motorized Transport integrated into Transportation Planning in Africa.* UK: Taylor Francis Ltd.

Morange, M., Folio, F., Peyroux, E. and Vivet, J. (2012) The spread of a transnational model: 'Gated communities' in three Southern African cities (Cape Town, Maputo and Windhoek). *International Journal of Urban and Regional Research* 36 (5), 890–914.

Mutongi, K. (2006) Thugs or Entrepreneurs? Perceptions of Matatu operators in Nairobi, 1970 to the present. *Africa: Journal of the International African Institute* 76 (4), 549–568.

Mutongi, K. (2017). *Matatu: A History of Popular Transportation in Nairobi.* Chicago: University of Chicago Press.

Myers, G. (2011) *African Cities: Alternative Visions of Urban Theory and Practice.* London: Zed Books.

Nizamo, H., Meyrowitsch, D.W., Zacarias, E. and Konradsen, F. (2006) Mortality due to injuries in Maputo City, Mozambique. *International Journal of Injury Control and Safety Promotion* 13 (1), 1–6.

Njoh, A. (2009) Urban planning as a tool of power and social control in colonial Africa. *Planning Perspectives* 24 (3), 301–317.

Olima, W. (1997) The conflicts, shortcomings, and implications of the urban land management system in Kenya. *Habitat International* 21 (3), 319–331.

Paasche, T. and James, S. (2010) Transecting security and space in Maputo. *Environment and Planning A* 42, 1555–1576.

Pieterse, E. (2010) *City Futures: Confronting the Crisis of Urban Development.* Cape Town: University of Cape Town Press.

Porter, G. (2007). Transport planning in sub-Saharan Africa. *Progress in Development Studies* 7 (3), 251–257.

Rasmussen, J. (2012) Outside the law – inside the system: Operating the Matatu sector in Nairobi. *Urban Forum* 23 (4), 415–433.

Rizzo, M. (2002) Being taken for a ride: Privatisation of the Dar es Salaam transport system 1983–1998. *Journal of Modern African Studies* 40 (1), 133–157.Rizzo, M. (2017) *Taken for a Ride: Grounding Neoliberalism, Precarious Labour and Public Transport in an African Metropolis*, Oxford: Oxford University Press.

Salon, D. and Gulyani, S. (2010) Mobility, poverty, and gender: travel 'choices' of slum residents in Nairobi Kenya. *Transport Reviews* 30 (5), 641–657.Schalekamp, H. and Behrens R. 2013. Engaging the paratransit sector in Cape Town on public transport reform: Progress, process, risks. *Research in Transportation Economics* 39 (1): 185–190.

Sclar, E., Lönnroth, M. and Wolmar, C. (2014) *Urban Access for the 21st Century: Finance and Governance Models for Transport Infrastructure*. New York, NY: Routledge.

Sietchping, R., Permezzi, J.M. and Ngomsi, C. (2012) Transport and mobility in sub-Saharan African cities: An overview of practices, lessons and options for improvements. *Cities* 29 (3), 183–189.

Todes, A., Karam, A., Klug, N. and Malaza, N. (2010) Beyond master planning? New approaches to spatial planning in Ekurhuleni, South Africa. *Habitat International* 34 (4), 414–420.

Townsend, A. (2013) *Smart Cities: Big Data, Civic Hackers, and the Quest for a New Utopia.* New York, NY: Norton and Co.

UN-Habitat (2013) *Planning and Design for Sustainable Urban Mobility: Global Report on Human Settlements*. Nairobi: UN-Habitat.

UN-Habitat and ITDP (2015) *Nairobi Ndovu/A104 BRT Service Plan*. Available at: www.itdp.org/wp-content/uploads/2015/02/Nairobi-Ndovu-A104-BRT-Service-Plan.pdf (accessed 20 March 2017).

Vanin, F. (2013) *Maputo, Open City*. Lisbon: Fundacao Serra Henriques.

Williams, S., Marcello, E. and Klopp, J.M. (2014) Open source Nairobi: Creating a GIS database for the city of Nairobi to provide equal access to information. *Annals of the Association of American Geographers* 104 (1), 114–130.

Williams, S., Waiganjo, P., White, A., Orwa, D. and Klopp, J. (2015) The digital Matatu project: Using cellphones to create open source data for Nairobi's semi-formal bus system. *Journal for Transport Geography* 49, 39–51.

Woolf, S.E. and Joubert, J.W. (2013) A people-centred view on paratransit in South Africa. *Cities* 35, 284–293.

World Health Organisation. (2013) *Road Safety in the WHO Africa Region: The Facts.* WHO. Available at: www.who.int/violence_injury_prevention/road_safety_status/2013/report/factsheet_afro.pdf (accessed 12 August 2016).

World Health Organisation. (2016) *Global Urban Ambient Air Pollution Database*. WHO. Available at: www.who.int/phe/health_topics/outdoorair/databases/cities/en/.

6 Exploring patterns of time-use allocation and immobility behaviours in the Bandung Metropolitan Area, Indonesia

Yusak O Susilo and Chengxi Liu

This chapter explores the patterns of time use and immobility behaviours in Bandung city, the second biggest metropolitan area in Indonesia. A three-week time-use and activity diary is used. The day-to-day variations in time use allocation across different socio-demographic groups are examined. The results show different distinct weekday and weekend patterns and mobile and immobile days' patterns of respondents' time use distribution. The results show a strong tendency of social exclusion resulting from transport poverty.

Introduction

Time use studies have interested sociology, geography, and transport researchers for decades. Whilst some disciplines focus on how time has been spent and distributed across different activities, and by different socio-demographic groups of population (e.g. Michelson, 2009; Papastefanou and Gruhler, 2014), other disciplines focus on the roles of one's time-space constraints in influencing one's time spend and travel and activity participation decisions (Hägerstrand, 1970; Merz et al., 2009; Susilo et al., 2012; Ellegård and Svedin, 2012) and wellbeing (e.g. Bonke and Gerstoft, 2007).

Most previous studies agree that different groups of populations have different patterns of time use, for different activities and trip purposes. The variability of the time use depends on the individual's socio-demographic characteristics and also his/her commitments with others (Susilo and Axhausen, 2014; Neutens et al., 2012; Liu et al., 2015a). Using data from nine European countries, Joesch and Spiess (2006) show that, on average, the population of those countries uniformly devoted almost 42 to 45% of the 24 hours of their time to physiological activities, 14% to family and house activities, 20% to public activities and 20% to personal activities. If one compares the time-budgets across gender and more detailed socio-demographic characteristics, however, it was evident that there are large differences between employment status and between men and women. In all observed countries, men devote more time to gainful work/study than to domestic work and men have overall more free time than women. Using the same data set, Fraire (2006) found that the countries that have similar cultural and geographical backgrounds are likely to have similar patterns of time use. For example, the British, Norwegian and Finish have relatively comparable time allocation patterns, whilst Estonian and Slovenian have different patterns of time use allocation.

In terms of time use for mobility and travel, there has been a long-standing notion of 'travel time budget' (TTB) – a fixed and stable amount of time that an individual can make available for travel (Susilo and Avineri, 2014). Some previous studies (e.g. Goodwin, 1973; Zahavi and Talvitie, 1980; Zahavi and Ryan, 1980; Newman and Kenworthy, 1999; Schafer and Victor, 2000) have observed the constancy and stability of travel time at the aggregated level, which is that, on average, an individual would spend about 1 to 1.5 hours per day for travelling. Some studies at the disaggregate level (e.g. Kitamura et al., 2006), however, have showed that travel routines and TTBs are not constant but rather are a function of several variables, which is different for different individuals and households.

It is reasonable to assume that these similarities and dissimilarities are related to unique conditions of individuals' time-space prisms (Hägerstrand, 1970). Whilst there is a relatively clear limit on the amount of time that an individual can spend on the given day (Gershuny et al., 1986; Ahmed and Stopher, 2014), people from various socio-demographic groups have different commitments, resources to use, and different external constraints, which subsequently results in their having different time allocation patterns (Susilo and Dijst, 2009, 2010). Some people have more stable activity and travel engagement patterns than others, whilst some others have more random patterns across days (Susilo and Axhausen, 2014; Heinen and Chatterjee, 2015). One may be immobile on one or two days, but travel farther on other days (Naess, 2006). The people who are immobile may not necessarily be less active than the ones who are mobile on the given day (Susilo and Liu, 2017). Furthermore, in many circumstances, this allocation of time is not only a result of a sole individual's decision, but also a result from intra- and inter-household interactions (Niemi, 2009; Habib and Miller, 2009; Chikaraishi et al., 2010; Arentze et al., 2011; Zhang et al., 2009; Susilo and Avineri, 2014; Liu et al., 2015a).

At the same time, immobility behaviour could also indicate an individual's non-participation in important activities and/or physical isolation at home because of his/her own constraints. Kenyon et al., (2002) explained that such exclusion can happen because of reduced accessibility to opportunities, services and social networks, caused by, in whole or in part, insufficient mobility in a society and environment built around the assumption of high mobility. Cass (2005) described various possible reasons of such conditions, including certain physical barriers that must be overcome to reach a certain destination by the traveller and also the financial cost that he/she needs to spend to enjoy the service. At the same time, Ureta (2008a) argued that such social exclusion does not necessarily mean an absolute immobility. Once a traveller would have devoted most of his/her motility resources (the capacity of the person to be mobile in social and geographical space) on mandatory trips, such as daily commuting, he/she may not afford/able to access and/or use the existing transport systems anymore (Kaufmann et al., 2004). In most cases, among travellers of lower-income groups, he/she would then have to cancel certain movements (trips) because the movement either became too expensive or because the trip was unnecessary given the available resources budget. Titheridge et al., (2014) and Lucas et al., (2016) summarised various evidences showing that poor transport and land use planning approaches contribute significantly to social exclusion by restricting access to activities that enhance people's life chances, such as work, study, health care, food, shopping and other important activities. Focusing on a specific group of travellers, Tacken (1998) and Delbosc and Currie (2011) found that a lack of available travel options can lead to harmful isolation that negatively influences well-being of vulnerable populations. Whilst some impact of such exclusion can be moderated with modern technologies, such as mobile phones, Ureta (2008b) argues that the impacts of such technology to enhance wellbeing among urban poor were found to be rather limited.

Further, it is important to remember that many findings about (im)mobility behaviours mentioned previously were based on trip/travel need–based analysis (based on the household travel survey data set or deep qualitative interview on one's travel pattern/travel needs), which defines that an immobile person is someone who has not declared any trips at all or is not able to make the trip; whilst in time-use literatures an immobile person can be defined either as someone who has remained at home the whole day or as someone with no travel activity, resulting from whatever reason. This different definition contributes to the discrepancy in the share of (im)mobility behaviours, ranging from 1.5 in Great Britain to 18 percentage points in France (Hubert et al., 2008). Because most of the previous studies have focused on the number of trips an individual made/desired every day, it is not clear yet what would be the impact of one's time-use allocations when he/she is not travelling, compared with the days of travelling. Furthermore, most of the previous studies on time-use allocation were based on cross-sectional data. Because an individual's needs and desires are not constant from day to day, an individual's travel pattern is neither totally repetitious nor new every day. Some activities (e.g. eating, sleeping) are repeated every day, whereas other activities such as shopping, personal business and social recreation are not necessarily repeated on a daily basis (Susilo and Axhausen, 2014; Dharmowijoyo et al., 2016a). This makes the distribution of time use a dynamic process, with learning and changing on the one hand and rhythms and routines on the other. Some activities are carried out daily, whilst others are carried out weekly or bi-weekly (Bayarma et al., 2007; Dharmowijoyo et al., 2017). Neglecting this day-to-day variability of an individual's time allocation and activity participation would lead to over/under-estimation of the influence of different key determinants towards one's time-use allocation and mobility behaviours.

In analysing such variability, however, it is also important to understand that whilst absolute clock time is a very efficient instrument to coordinate the time-space paths of humans, artefacts and other objects because of the homogeneous and context-independent conception that it enables, in many cases, humans cannot fully grasp synchronisation processes by concentrating on clock time alone (Schwanen, 2006). Many previous studies (Cullen and Godson, 1975; Kim and Kwan, 2003) have shown that an assumption that punctuality and synchronisation are something that means absolute for every actor involving at any given time is not necessarily correct. To explain this phenomenon, Schwanen (2006) gave an example that a mother can consider herself to be 'too late' in a relational sense to pick up her child at nursery but 'on time' according to the nursery's clock time. This happens when the mother perceives her opportunities for combining work and domestic responsibilities to be more limited, than a focus on the clock-based opening time of the childcare centre in relation to her working hours.

Most previous studies were based on cases in developed countries. In many respects, time allocations and mobility patterns within a household are intertwined with one's cultures, expectations and habits. Thus, different cultures would have various mechanisms for allocating responsibility, trip and time allowance within households. Some developing countries, such as Indonesia, have been growing very fast in the past decade. There enormous changes have occurred in the physical and social environments of trip making in a short time, which makes us argue that variations in the observed amount of time allocated for an individual's travel are becoming more complex and less predictable (Susilo and Joewono, 2017). Various new commodities, appliances and services have been introduced to reduce the time required for domestic chores such as cleaning, cooking and gardening. In more educated families, two-worker households have become a social norm rather than an exception, changing the way in which household tasks are carried out by its members. All of these changes affect the needs for, resources available for and constraints imposed on travel (Susilo and Kitamura, 2008). These changes, however, may variously affect different societal groups. Some may adopt the changes faster whilst others may not experience change. The different adaptation abilities between rich and poor residents to follow and adopt these changes may widen the economic segregation and also mobility opportunities between these groups.

To understand this phenomenon, in a multiday context, this chapter aims to explore the patterns of time use and immobility behaviour in the city of Bandung, Indonesia, across different socio-demographic groups. A three-week time-use and activity travel diary is used for this purpose. The differences of time use and immobility among household members are explored and discussed. Time-use patterns among people of different socio-demographic groups during the various days of the week are investigated, providing evidence of significant impact of immobile behaviours on individuals' time use, and thus implying different travel constraints, resources and opportunities for individuals who fill different roles in family and society.

In the following section, the study area and the three-week time-use and activity travel diary are introduced. Four time-use categories are defined. Following this is a section presenting a descriptive analysis of individuals' time-use allocations on different days and across various socio-demographic groups. The chapter closes with a discussion section.

The study area and the description of the data set

The Bandung Metropolitan Area (BMA)

The Bandung Metropolitan Area (BMA) is the capital of the province of West Java and is approximately 200 km or two to three hours' drive south of Jakarta. With its conurbation, the BMA population is approximately 7.89 million people; they live in an area that covers 3,382.89 square km. The BMA is the second largest metropolitan area in Indonesia after the Jakarta Metropolitan Area. A typical city in developing countries, the BMA has a very relaxed or unplanned mixed and monocentric land use, congested road networks and poor public transport networks and services (Susilo et al, 2011). Road congestion and the low performance of public transport/paratransit services encourage travellers in the BMA to use motorcycles to reduce their costs and time (Susilo, 2011; Tarigan et al., 2014; Susilo et al, 2015). At the same time, they usually have more choices within a closer range in which to conduct their activities along their travel routes, because of the highly mixed land-use configurations (Dharmowijoyo et al., 2014, 2016b).

The 2013 BMA data set

The BMA data set includes data covering household, physical activity and lifestyle, individual's subjective characteristics, time-use and activity diary and subjective wellbeing. The survey involved 732 individuals and 191 households from all over the BMA for 21 consecutive days. The household data section contains household composition, individuals' perceptions about how far his/her accommodations are from the city centre, public and transportation facilities, and build environment variables. Time-use and activity diary surveys captured 23 in-home and out-of-home activity classifications, travel duration and mode characteristics, and multi-tasking activities for adults, young adults and children older than age 7. The locations of all in-home and out-of-home activities were geo-coded. The time-use activity diary survey was sampled at 15-minute intervals (i.e. 96 time slices in one day). This method caused less bias in estimating the time spent allocated to the respective types of activities and was easily operationalised by surveyors and respondents in the cases of developing countries. However, this method failed to record short-duration activities below 15 minutes (Dharmowijoyo et al., 2015). In this survey design, time-use activity participation was classified into 23 different activities types, including travel. Those activity types are further classified into four distinct activity time categories based on the nature of activity time use (Hägerstrand, 1970; Schwanen et al., 2008; Akar et al., 2011; Ellegård and Svedin, 2012; Susilo and Axhausen, 2014). Following are the definitions of those four distinct activity time categories, and classifications are presented in Table 6.1.

Contracted time: Contracted time refers to the time a person allocates towards an agreement to work or study. When a person is using contracted time to commute, this person understands that this travel time is directly related to paid work or study and any break in this commute time directly affects job- or school-related performance.

Committed time: Committed time, like contracted time, takes priority over necessary and free time because it is viewed as productive work. It refers to the time allocated to maintain a home and family. When a person is commuting using committed time, this person may feel that the commute is a duty to family, such as walking children to school or driving a spouse to work. Contracted and committed time users may feel that their commute is more important than the commute of necessary or free time users because their commute is productive work. Therefore, they may be more inclined to choose a motorized mode of travel.

Necessary time: Necessary time refers to the time required to maintain one's self as it applies to activities such as eating, sleeping and cleansing and to a large

Table 6.1 Classification of activity categories

Activity categories	*Original activity classification in the survey*
Contracted time	Work
	School
Committed time	Household activities such as cleaning the house, cooking or baking, washing clothes/dishes, etc.
	Babysitting activities, including babysitting, playing with the baby, feeding the baby, etc.
	Selling and purchasing activities
	Daily grocery shopping
	Picking up/dropping off children
Necessary time	Sleeping
	Personal care activities, such as taking a bath, brushing teeth, etc.
	Eating/drinking at home
	Organisation/volunteer/political activities such as youth/political/religious meetings, visiting mosques, etc.
	Maintenance activities, including going to hospital/health centre/medical doctor, visiting bank/post office
	Fixing mechanics, such as going to a mechanic shop
Free time	Relaxing activities such as watching TV, listening to radio, reading newspapers, relaxing, etc.
	Social/family activities such as chatting with family members, visiting friends, etc.
	Eating/drinking outside such as eating in a restaurant
	Sports activities such as going to a gym, playing football, etc.
	Holidays

extent, exercising. People who commute using necessary time may feel that the commute is an important activity for personal wellbeing and may also take into account the wellbeing of the natural and social environment. The person commuting in necessary time may be more inclined to choose an active mode of transportation for personal reasons that include exercise on top of transportation. Because sleeping is included in this category, necessary time usually constitutes the majority of people's time.

Free time: Free time refers to the remains of the 24-hour day after the three other types of time have been subtracted from it. This type of time is not necessarily discretionary time as the term "free" time may imply, because people tend to plan activities in advance, creating committed free time in lieu of discretionary time. People who commute using free time are more apt to view the commute as a recreational activity. Commuting in free time provides the greatest gains for social capital because the person commuting in free time is more likely to slow down or stop the commute at his discretion to undertake another activity or engage in social interaction. He or she may also view the commute as part of destination activity to which he or she has gladly committed free time.

It is worth noting that a given individual can do several activities at the same time, for example, taking care of a baby whilst at the same time eating at home.

Those multi-tasking activities are defined as concurrent activities containing primary and secondary activities (Kenyon, 2010; Circella et al, 2012) for satisfying different needs and desires at the same time. More detailed descriptions about the data set and survey design can be found at Dharmowijoyo et al., (2015). The profiles of the respondents and the distribution of time allocation across different activities is presented in Table 6.2.

As depicted in Table 6.2, the sample consists of a balanced distribution of socio-demographic profiles. Most of the respondents (75%) come from low-income

Table 6.2 Sample profiles

Socio-demographic and travel activity characteristics	*Percentage/mean*
Socio-demographic characteristics at individual level:	
Male	54.41%
Worker	43.64%
Non-worker	31.05%
Student	25.31%
Aged 15–22 years	18.60%
Aged 23–45 years	44.76%
Aged 46–55 years	14.27%
Older than 55 years of age	8.81%
Part of low-income households (< IDR 3 million/month)	75.39%
Part of medium-income households (IDR 3–6 million/month)	16.23%
Part of high-income households (>IDR 3–6 million/month)	8.38%
Household characteristics:	
Number of household members	3.97
Number of dependent children per household	1.56
Number of motorised vehicles per household	1.8
Reside within the CBD of BMA	17.20%
Reside within the inner city boundary of BMA	44.90%
Reside within Greater BMA region	37.90%
Trip engagements and travel time spent across different days of the week:	
Number of trips/day	2.49
Number of trip chains/day	1.18
Percentage of using motorised mode on the given day	39.05%
Percentage of using public transport mode on the given day	13.49%
Percentage of using non-motorised mode on the given day	33.26%
Average total travel time spent/day (min, % of time spent on the given day)	72.91 (5.07%)*
Contracted time spent per day (min, % of time spent on the given day)	194.22 (13.49%)*
Committed time spent per day (min, % of time spent on the given day)	159.56 (11.08%)*
Necessary time spent per day (min, % of time spent on the given day)	615.52 (42.75%)*
Free time spent per day (min, % of time spent on the given day)	456.71 (31.72%)*
Time engaging with multi-tasking activities jointly with other activities: (min, % of time spent on the given day)	

(*Continued*)

Table 6.2 (Continued)

Socio-demographic and travel activity characteristics	*Percentage/mean*
Time engaging with multi-tasking activities within contracted time activities	24.23 (12.26%)*
Time engaging with multi-tasking activities within committed time activities	36.26 (22.77%)*
Time engaging with multi-tasking activities within necessary time activities	25.92 (4.21%)
Time engaging with multi-tasking activities within free time activities	163.58 (35.95%)*
Time engaging with multi-tasking activities during travel	10.92 (14.35%)*

Note: Trip chain is defined as home-to-home trip. For many workplaces, Saturday is still a half workday.

*The percentages in brackets show the average proportions of time spent for travel and for each type of in-home and out-of-home activity.

IDR, Indonesian rupiah.

households. The sampled households on average have four household members with more than one child in the household. It is a tradition in Indonesia that elders live together with their adult children, thus leading to a common phenomenon of households with more than four members. Although only a small share of sampled households own a car, most sampled households own at least one motorcycle; the motorcycle is the major motorised mode used in Indonesia. It is also worth noting that only 20% of the respondents live in CBD (central business districts) of urban zones while a large share of respondents live in suburban and rural areas where housing prices are cheaper. However, the highly mixed land use pattern does not necessarily ensure CBD as a more accessible area for various activities such as grocery shopping, compared with a non-CBD area.

In terms of general travel pattern variables, an average respondent makes 2.5 trips per day. Average number of trips per trip chain is two, indicating that most respondents conduct simple home-based, go-and-back trips. In terms of mode share, approximately 40% of total trips travelled per day are made by motorised modes. Public transport modes comprise only 14% of total trips made. Total travel time spent reaches 74 minutes per day.

In terms of time-use patterns on different days of the week, the average contract time is 194 minutes. The average contract time from Monday through Thursday of those who are working and studying is 404 minutes, which is slightly lower than the number in Sweden (445 minutes) (Liu et al., 2015b). This is presumably because the sample contains a relatively low share of permanent workers (38.1%) among those who are working and studying on the given day but a considerable share of temporal workers (23.3%) and students (22.3%), who are believed to have shorter contracted time per day than permanent workers.

On average, 12% of total contracted time is multi-tasking, while 22% of committed time is multi-tasking. Only 4% of necessary time is multi-tasking since most necessary time is sleeping time. Free time has the highest share of multi-tasking – 35% – since free time is the least physically and mentally intensive. A total of 14% of travel time is multi-tasking, which is mainly travel time by public transport and walking. It is also worth noting that the multi-tasking shares in all types of activity time use do not differ much between weekday and weekend (specific numbers are not shown in this paper). This finding seems to suggest a consistent multi-tasking activity pattern for a given activity time use, for example, contracted time use, even when one may spend much longer or shorter times on that activity type on the weekend compared with on a weekday.

Descriptive analyses: the distribution of time use across days and across different group of population in the BMA

Time-use allocation patterns and immobile behaviours across days and across different individual socio-demographic groups are explored in order to further understand the day-to-day variability and between-individual heterogeneity in time use and mobility pattern.

Day-to-day variability of activity time allocation

Figure 6.1 presents the day-to-day variability of activity time allocation. As presented in Figure 6.1, time use from Monday through Thursday is quite stable. Free time increases by around 20 minutes on Friday compared with Monday through Thursday, while a similar amount of decrease in contracted time is also observed. The time-use distribution shows a gradual but considerable change from Friday through Sunday. The time-use distribution on Sunday shows a clear difference compared with Friday and even to Saturday. Contracted time on Sunday decreases by 140 minutes compared with Friday and 100 minutes compared

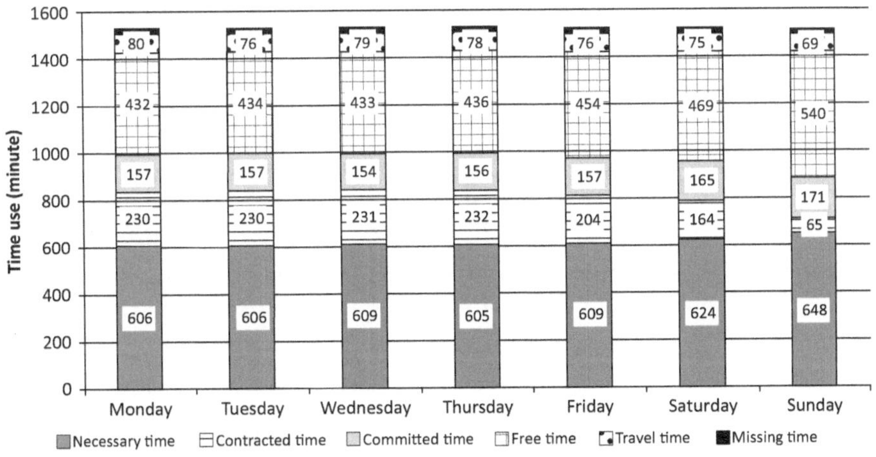

Figure 6.1 Day-to-day time use comparison

with Saturday. The contracted time on weekdays is redistributed mostly into free time and partially into necessary time on weekends.

Table 6.3 a and b present the time-use distribution across individual socio-demographic groups, including income level, gender and occupation.

As shown in Table 6.3, respondents from the high-income group in general have a shorter contracted time and committed time but longer necessary time on weekdays compared with respondents from the medium- and low-income groups. Respondents from the high-income group may not need to work as many hours as those from the low-income group, but may have more maintenance activities and personal care activities because they can afford more personal properties which then generate maintenance errands. Travel time on weekdays of respondents from the low-income group is significantly shorter than that on weekdays of those from the medium- and high-income groups. Presumably, those from the low-income group may only travel within vicinity, and thus have a shorter travel time because of limited travel resources (such as car availability and low travel budget). During the weekend, respondents from the high-income group also show a shorter contracted time and committed time. However, unlike the pattern shown on weekdays, respondents from the high-income group have a longer free time

Table 6.3 (a) Weekday time-use distribution (min) among individuals in different socio-demographic groups

		Necessary time	*Contracted time*	*Committed time*	*Free time*	*Travel time*	*Missing time*
Household income	Low income	606.9***	217.6***	166.3**	440.8	71.8***	27.8***
	Medium income	595.9***	230.8***	169.9**	443.3	95.7***	10.7***
	High income	620.7***	190.4***	147.1**	438.7	92.5***	23.6***
Gender	Male	594.1***	295.1***	77.7***	442.7***	98.2***	20.8***
	Female	622.8***	146.7***	246.8***	431.2***	54.3***	27.3***
Occupation	Permanent worker	581.0***	329.3***	89.3***	400.7***	106.9***	31.1***
	Temporal worker	585.4***	355.8***	94.6***	366.0***	88.9***	29.3***
	Part-time worker	598.4***	187.0***	205.9***	380.9***	78.6***	67.5***
	Non-worker	681.3***	81.3***	191.2***	528.9***	45.1***	10.1***
	Student	621.9***	297.0***	48.2***	438.7***	89.4***	3.5***
	Household wife	626.8***	44.3***	324.6***	486.6***	38.6***	14.8***
	Retired	692.3***	33.5***	167.4***	612.5***	59.7***	1.1***
	Others	554.5***	194.6***	174.2***	494.0***	77.7***	48.6***

Note: *** denotes that the corresponding time category is significantly different between socio-demographic groups at the 1% level. ** denotes that the corresponding time category is significantly different between socio-demographic groups at the 5% level. * denotes that the corresponding time category is significantly different between socio-demographic groups at the 10% level.

Table 6.3 (b) Weekend time-use distribution (min) among individuals in different socio-demographic groups

		Necessary time	*Contracted time*	*Committed time*	*Free time*	*Travel time*	*Missing time*
Household income	Low income	637.2	111.1***	179.0***	497.7***	66.6***	25.1***
	Medium income	627.5	112.3***	187.3***	512.4***	80.4***	11.5***
	High income	635.4	94.2***	143.7***	529.4***	90.4***	10.2***
Gender	Male	628.4***	152.7***	89.3***	534.2***	86.7***	20.0
	Female	645.1***	71.1***	258.5***	469.5***	54.8***	22.0
Occupation	Permanent worker	630.1***	165.4***	117.0***	501.4***	83.5***	23.6***
	Temporal worker	616.6***	159.3***	140.6***	486.1***	80.9***	26.1***
	Part-time worker	618.9***	129.5***	205.2***	405.9***	75.9***	72.1***
	Non-worker	676.6***	70.5***	185.4***	543.9***	49.0***	7.9***
	Student	673.8***	126.3***	66.0***	545.2***	82.2***	2.8***
	Household wife	633.4***	23.1***	312.8***	492.1***	48.0***	13.1***
	Retired	687.2***	26.5***	152.8***	625.8***	74.0***	0.3***
	Others	569.1***	167.7***	128.8***	506.3***	72.1***	47.7***

on weekends than those from the medium- and low-income groups, but necessary time use is comparable among income groups on weekends. On weekends, those from the high-income group are more capable and able to afford to conduct leisure activities compared with those from the medium- and low-income groups because of having sufficient available time and money, and thus are able to have a longer free time.

Women have shorter contract, free and travel times but longer necessary and committed time than men on both weekdays and weekends. Women, often as housewives in traditional households in Indonesia, are responsible for housework and maintenance activities, and therefore have longer necessary and committed times.

In terms of occupation type, permanent and temporal workers in general have much shorter committed and necessary times but slightly longer travel time compared with non-workers, household wives and retirees. Household wives have a particularly high committed time, as expected, but a particularly low travel time, indicating that housewives may not travel far distances. Retirees and non-workers have the longest necessary time and free time among all categories. Retirees may need more sleeping time because they are elders, while non-workers may also have longer personal care activities and maintenance activities since they do not have tight time constraints compared with workers. To examine the time-use

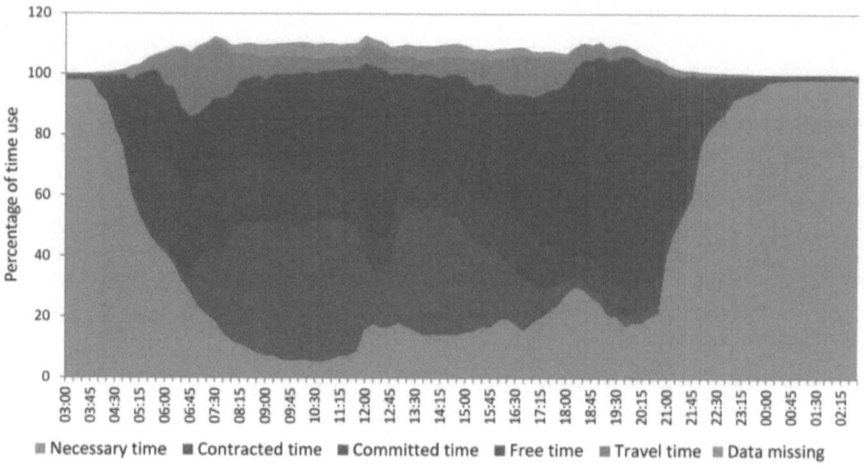

Figure 6.2 (a) Daily time use distribution on weekdays

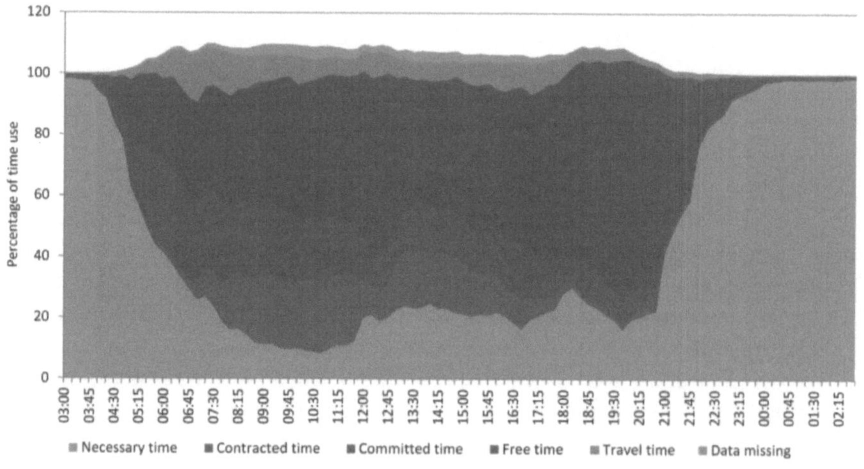

Figure 6.2 (b) Daily time-use distribution on weekends

distribution within a day, the daily time-use distributions across 24 hours on weekdays and weekends are presented in Figure 6.2.

As shown in Table 6.2, time use can exceed 100% as a result of multi-tasking activities. The multi-tasking activities consume approximagely 10% of total time use from 5:00 to 21:00 on both weekdays and weekends. It is shown clearly in the weekday time-use distribution (Figure 6.2 a) that the morning peak of travel is from 6:00 to 9:00. Before 6:00, from 4:30 to 6:00, the contracted time and free time increase with a decrease of necessary time, as people get up and start doing personal care activities and household activities. From 6:45, the contracted time expands, indicating that people arrive at the work place and start working. A peak of free time is observed at 12:00, showing that most people have lunch at this time. The afternoon travel peak is much smoother than the one of morning peak, meaning that workers leave work at different times. Correspondingly, free time after 17:00 increases dramatically, meaning that people start their after-work relaxing. Free time reaches its peak at 20:00 after dinnertime at 18:30 (see the peak of necessary time at 18:30), and then necessary time increases sharply, signifying the time of going to bed.

The weekend time-use distribution shows a similar pattern but with a much lower share of contracted time and much smoother morning peak of travel. Free time on weekends is much longer than that on weekdays, while committed time on weekends is also slightly longer than that on weekdays.

The interaction between individual's socio-demographics and travel time, mobility patterns

Table 6.4 presents the immobile behaviour and travel time per mobile day across different socio-demographic groups. Respondents from the low-income group have the most immobile days during the three-week survey period among income groups, four percentage points more immobile days compared with those from

Table 6.4 Immobile days and average travel time per day across socio-demographic groups

		Percentage of immobile days		*Average travel time per mobile day*	
		Mean (%)	*S.D.*	*Mean (min)*	*S.D.*
Household income	Low income	21.41*	0.30	85.4***	48.1
	Medium income	19.10*	0.26	106.3***	68.6
	High income	17.70*	0.30	90.5***	53.2
Gender	Male	11.64***	0.21	105.4***	59.4
	Female	29.35***	0.34	72.7***	37.6
Occupation	Permanent worker	9.68***	0.19	107.7***	61.4
	Temporal worker	12.00***	0.19	96.9***	51.5
	Part-time worker	17.41***	0.29	88.7***	53.9
	Non-worker	38.69***	0.37	68.7***	40.5
	Student	10.44***	0.18	97.7***	39.2
	Household wife	37.79***	0.36	63.3***	36.2
	Retired	40.91***	0.37	100.9***	77.1
	Others	16.26***	0.28	88.8***	53.7

Note: *** denotes that the corresponding time category is significantly different between socio-demographic groups at the 1% level. ** denotes that the corresponding time category is significantly different between socio-demographic groups at the 5% level. * denotes that the corresponding time category is significantly different between socio-demographic groups at the 10% level.

the high-income group. Respondents from the low-income group also have the shortest travel time per mobile day among income groups, indicating that those from the low-income group mostly travel within vicinity. Females, as expected, are much more immobile and have shorter travel times per mobile day than males. Among different occupation groups, non-workers, household wives and retirees have high percentages of immobile days compared with workers and students. However, retirees have much longer travel times per mobile day compared with household wives and non-workers, which may imply that retirees, although limited by their physical conditions, also have the need to travel (Andrews et al., 2012; Musselwhite et al., 2015). Household wives have the shortest travel time per mobile day. Students' travel time per mobile day is long (97.7 min) but with a relative small standard deviation (39.2 min), showing that all sampled students have similar travel time per mobile day, most ranging from 60 minutes to 120 minutes.

Figure 6.3 a and b present the immobile behaviour and travel time per mobile day across income groups between weekdays and weekends. It is not surprising that the percentage of immobile days on weekends is much higher than that on weekdays. The low-income group has the highest percentage of immobile days on weekdays, while the medium-income group has the highest percentage of immobile days on weekends. People from the medium-income group are likely to be permanent/temporal workers who have a regular work pattern, working on weekdays and resting on weekends. In terms of travel time per mobile day, the medium-income group has the longest travel time per mobile day on weekdays because of the commute travel time. However, on weekends, the high-income group has the longest travel time per mobile day because of their affordability of cars and weekend leisure travels.

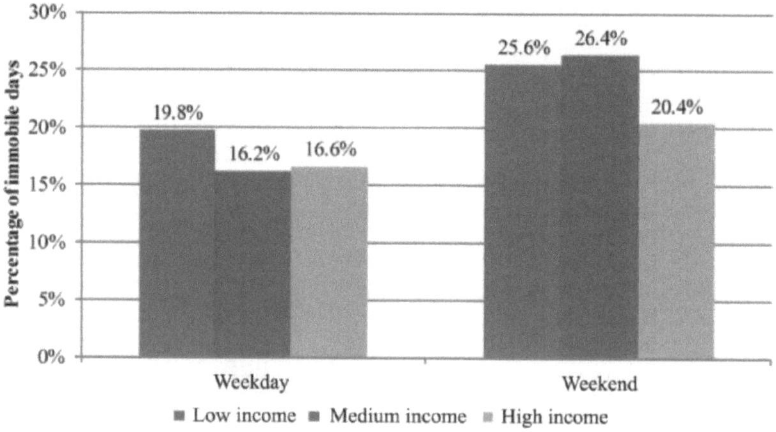

Figure 6.3 (a) Percentage of immobile days

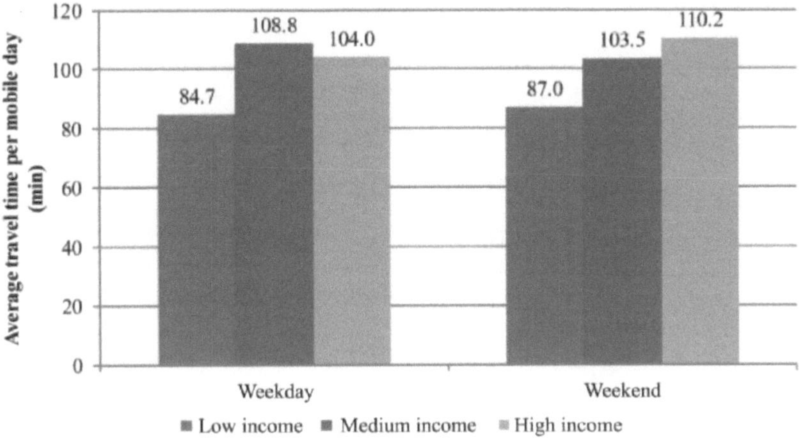

Figure 6.3 (b) Percentage of travel time per mobile day

The time-use comparison between mobile days and immobile days of individuals from three income groups, during weekday and weekend days, are presented in Figure 6.4. For individuals of low income, a strong correlation is observed between work time and mobile/immobile behaviour, indicating that being off work is a chief reason for being immobile. However, those of low income spend more time on sleeping, social activities and relaxing activities (see the activity classification in Table 6.1) on immobile days than on mobile days. This trend seems to be consistent between weekdays and weekends. A plausible interpretation of this trend could be that people of low income are under budget constraints and thus have limited options in travel; hence, they are more likely to choose to spend time on relaxing, social activities and sleeping at home. It is worth noting that those of low income also have longer times multi-tasking on immobile days than on mobile days, particularly on the weekend. This could indicate that household members in low-income families share in-home housework on weekends. Similar patterns are also observed for individuals of medium income and of high income. Those of medium income have longer times multi-tasking on weekdays than weekends (the opposite trend compared with the low-income group), indicating that the medium-income group is often under tight time constraints during weekdays. They also spend much longer times on household activities, social and eating outside on weekdays when they are immobile than on weekends. For those of high income, the difference between mobile and immobile days is comparable between weekdays and weekends, except that they spend less time relaxing and more time doing social and household activities when they are immobile on weekdays than on weekends.

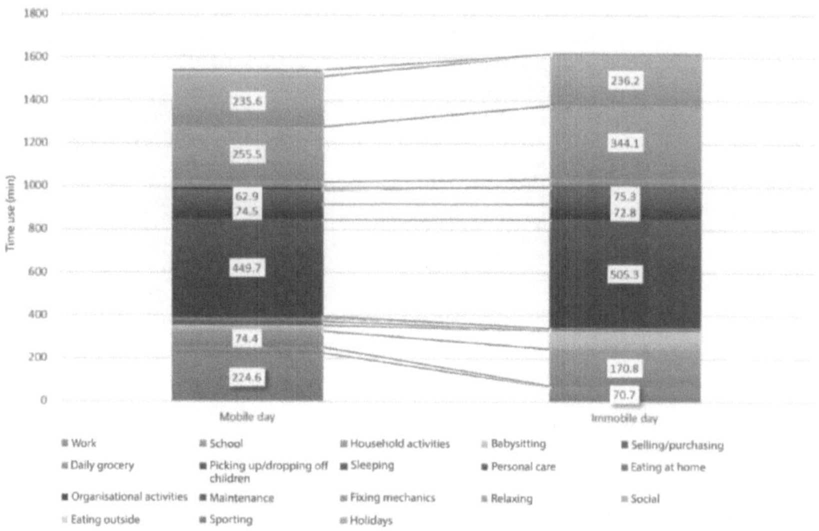

Figure 6.4 (a) Weekday time use of the low-income group

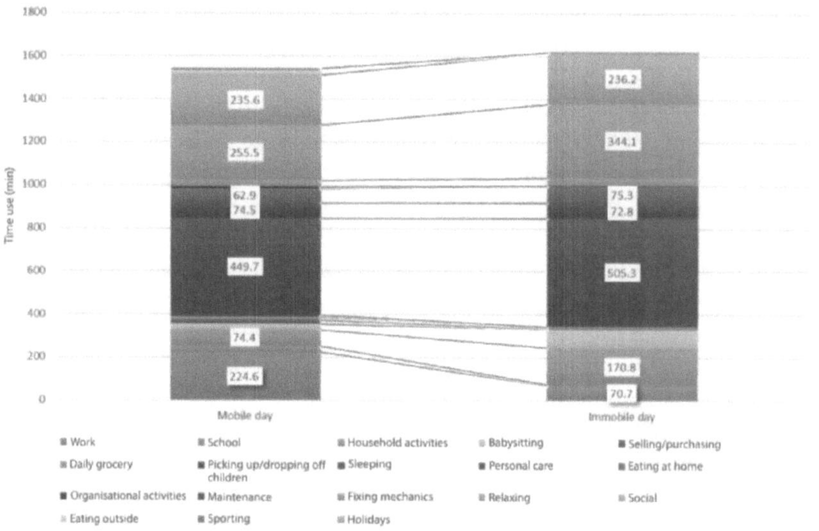

Figure 6.4 (b) Weekend time use of the low-income group

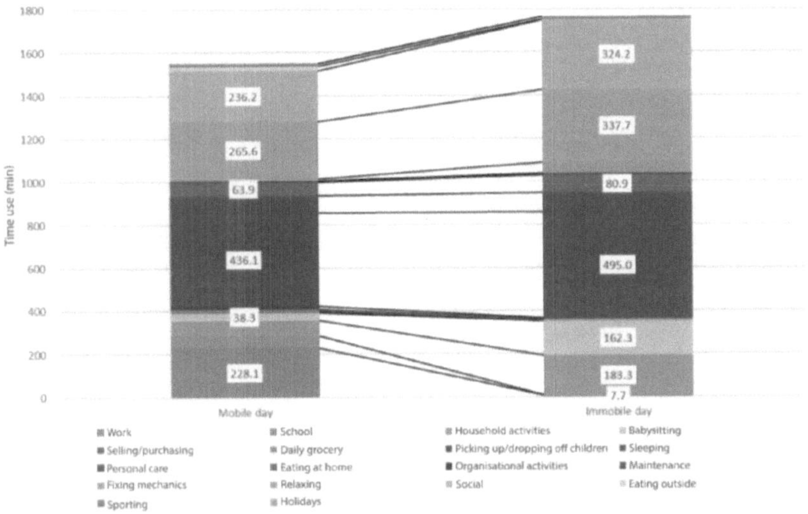

Figure 6.4 (c) Weekday time use of the medium-income group

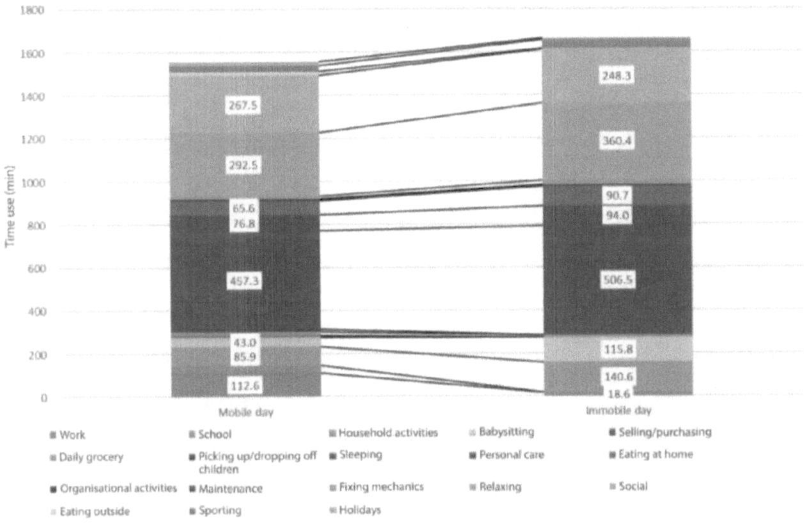

Figure 6.4 (d) Weekend time use of the medium-income group

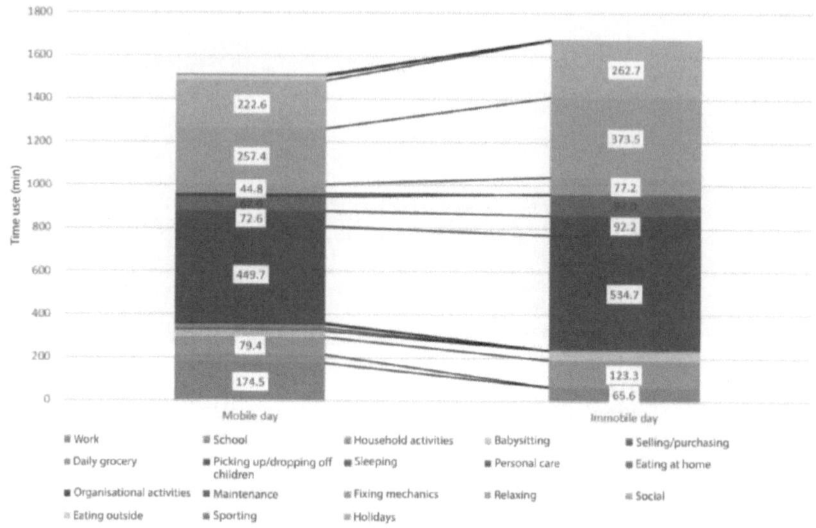

Figure 6.4 (e) Weekday time use of the high-income group

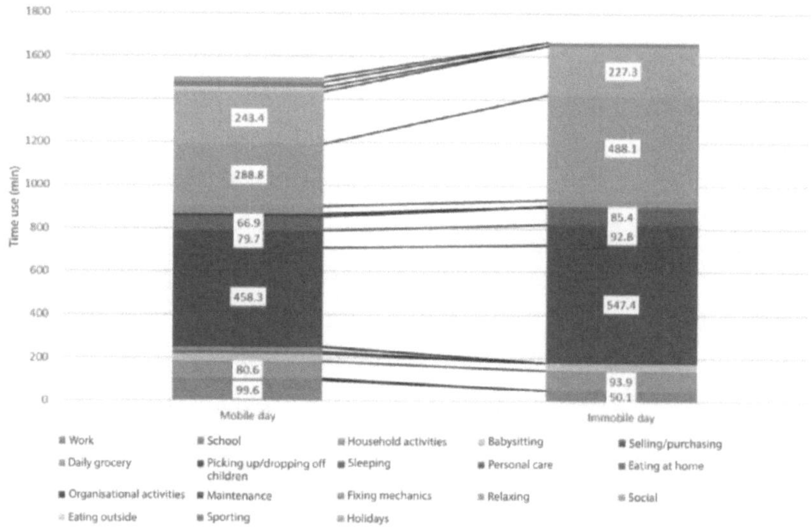

Figure 6.4 (f) Weekend time use of the high-income group

Conclusion

This chapter explores the time-use allocation and immobile behaviour using a three-week time-use diary collected in Bandung Metropolitan Area (BMA), Indonesia, which is one of the first multi-day time-use studies undertaken within the developing countries context. In line with findings in developed countries, there were clear differences between weekday and weekend patterns of time-use allocations and mobility behaviours across individuals from different socio-demographic groups.

The high-income group, however, was not found to be the one that works the most. On the contrary, the medium- and low-income groups are the ones who spend the most time on contracted and committed activities on weekdays. At the same time, on weekends, the medium- and low-income groups are less mobile than the high-income group. This is in line with Naess (2006) and Susilo and Avineri (2014), who argue that low- and medium-income individuals have a limited time and resources budget to spend. Thus, once one has worked and commuted extensively on weekdays, one will need to rest on weekends and stay at home. This can be confirmed from the trend showing that, whilst respondents from the high-income group spend more free time during weekends, the medium-income group has the highest percentage of immobile days on weekends.

That said, there is still a strong tendency of social exclusion because of transport poverty. Even on weekdays, the low-income group has the highest percentage of immobile days. Whilst those of low income spend more time on sleeping, social activities and relaxing activities on immobile days, those of high income spend more time for doing social and household activities. When they are mobile, the high-income group has the longest travel time per mobile day because they can afford cars and weekend leisure travels.

As a typical patriarchal society in a developing country, women are still responsible for housework and thus have shorter contract, free and travel times but longer necessary and committed times than men on both weekdays and weekends. Housewives also have a particularly low travel time, indicating that household wives may not travel far distances. Retirees and non-workers have the longest necessary and free times. Retirees may need more sleeping time as they are elders, while non-workers may also have longer personal care and maintenance activities since they do not have tight time constraints compared with workers.

It is important to remember that this study has employed multivariate analysis to systematically analyse and quantitatively quantify the impacts of each variable towards the immobility behaviours and time-use allocation in the BMA. Furthermore, this study, although it includes variables such as number of household members, does not model intra-household member interaction. Because fathers, mothers and children play different roles in households, their travel behaviour and time use are also intertwined with one another. Those are possible future research directions. Further model development could focus on modelling time use and immobile behaviour of each household member in an integrated model system.

References

Ahmed, A. and Stopher, P.R. (2014). Seventy minutes plus or minus 10 – a review of travel time budget studies. *Transport Reviews* 34 (5), 607–625.

Akar, G., Clifton, K.J., Doherty, S.T. (2011) Discretionary activity location choice: In-home or out-of-home. *Transportation* 38, 101–122.

Andrews, G., Parkhurst, G., Susilo, Y.O. and Shaw, J. (2012) 'The Grey Escape': How and why are older people using their free bus passes? *Journal of Transport Planning and Technology* 35 (1), 3–15.

Arentze, T.A., Ettema, D. and Timmermans, H.J.P. (2011) Estimating a model of dynamic activity generation based on one-day observations: Method and results. *Transportation Research Part B* 45, 447–460.

Bayarma, A., Kitamura, R. and Susilo, Y.O. (2007) On the recurrence of daily travel patterns: A stochastic-process approach to multi-day travel behavior. *Transportation Research Record* 2021, 55–63.

Bonke, J. and Gerstoft, F. (2007) Stress, time use and gender. *International Journal of Time Use Research* 4 (1), 47–68. dx.doi.org/10.13085/eIJTUR.4.1.47–68.

Cass, N., Shove, E. and Urry, J. (2005). Social exclusion, mobility and access. *The Sociological Review* 53 (3), 539–555.

Chikaraishi, M., Zhang, J., Fujiwara, A. and Axhausen, K.W. (2010) Exploring variation properties of time use behavior on the basis of a multilevel multiple discrete-continuous extreme value model. *Transportation Research Record* 2156, 101–110.

Circella, G., Mokhtarian, P.L. and Poff, L.K. (2012) A conceptual typology of multitasking behaviour and polichronicity preferences. *International Journal of Time Use Research* 9, 59–107.

Cullen, I. and Godson, V. (1975) Urban networks: the structure of activity patterns. *Progress in Planning* 4, 1–96.

Delbosc, A. and Currie, G. (2011) The spatial context of transport disadvantage, social exclusion and well-being. *Journal of Transport Geography* 19, 1130–1137.

Dharmowijoyo, D.B.E., Susilo, Y.O. and Karlström, A. (2014) The day-to-day inter and intra personal variability of individual's activity space in developing country. *Environment and Planning B: Planning and Design* 41, 1063–1076.

Dharmowijoyo, D.B.E., Susilo, Y.O. and Karlström, A. (2016a). On complexity and variability of individuals' day-to-day discretionary activities. *Transportation*. doi: 10.1007/s11116-016-9731-5.

Dharmowijoyo, D.B.E., Susilo, Y.O. and Karlström, A. (2016b) The day-to-day variability in travellers' activity-travel patterns in the Jakarta metropolitan area. *Transportation* 43, 601–621. doi: 10.1007/s11116-015-9591-4.

Dharmowijoyo, D.B.E., Susilo, Y.O. and Karlström, A. (2017) Analysing the complexity of day-to-day individual activity-travel patterns using a multidimensional sequence alignment model: A case study in the Bandung metropolitan area, Indonesia. *Journal of Transport Geography*, 64, 1–12.

Dharmowijoyo, D.B.E., Susilo, Y.O., Karlström, A. and Adiredja, L.S. (2015) Incorporating three-weeks' household time-use and activity diary with individual attitudes, physical activities and psychological characteristics in the Bandung metropolitan area. *Transportation Research Part A* 80, 231–246.

Ellegård, K. and Svedin, U. (2012) Torsten hägerstrand's time-geography as the cradle of the activity approach in transport geography. *Journal of Transport Geography* 23, 17–25.

Fraire, M. (2006) Multiway data analysis for comparing time use in different countries – application to time budgets at different stages of life in six European countries. *International Journal of Time Use Research* 3 (1), 88–109, dx.doi.org/10.13085/eIJTUR.3.1.88–109.

Gershuny, J., Miles, I., Jones, S., Mullings, C., Thomas, G. and Wyatt, S. (1986) Time budgets: Preliminary analyses of a national survey. *The Quarterly Journal of Social Affairs* 2 (1), 13–39.

Goodwin, P. (1973) *Time, Distance and Cost of Travel by Different Modes*. A Paper presented at Universities' Transport Study Group (UTSG) Conference, UCL, London.

Habib, K.M.N. and Miller, E.J. (2009) Reference-dependent residential location choice model within a relocation context. *Transportation Research Record* 2133, 92–99.

Hägerstrand, T. (1970) What about people in regional science? *Papers of the Regional Science Association* 24, 7–21.

Heinen, E. and Chatterjee, K. (2015) The same mode again? An exploration of mode choice variability in Great Britain using the National Travel Survey. *Transportation Research Part A* 78, 266–282.

Hubert, J.P., Armoogum, J., Axhausen, K.W. and Madre, J.L. (2008). Immobility and mobility seen through trip-based versus time-use surveys. *Transport Reviews* 28 (5), 641–658.

Joesch, J.M. and Spiess, C.K. (2006) European mothers' time spent looking after children – differences and similarities across-nine countries. *International Journal of Time Use Research* 3 (1), 1–27. dx.doi.org/10.13085/eIJTUR.3.1.1–27.

Kaufmann, V., Bergman, M.M. and Joye, D. (2004) Motility: mobility as capital. International Journal of Urban and Regional Research, 28, 745–756.

Kenyon, S. (2010) What do we mean by multitasking? Exploring the need for methodological clarification in time-use research. *International Journal of Time Use Research* 7, 609–619.

Kenyon, S., Rafferty, J. and Lyons, G. (2002) Transport and social exclusion: Investigating the possibility of promoting inclusion through virtual mobility. *Journal of Transport Geography* 10 (3), 207–219.

Kim, H.M. and Kwan, M.P. (2003) Space-time accessibility measures: A geocomputational algorithm with a focus on the feasible opportunity set and possible activity duration. *Journal of Geographical Systems* 5, 71–91.

Kitamura, R., Yamamoto, T., Susilo, Y.O. and Axhausen, K.W. (2006) How routine is a routine? An analysis of the day-to-day variability in prism vertex location. *Transportation Research part. A* 40, 259–279.

Liu, C., Susilo, Y.O. and Dharmowijoyo, D.B.E. (2015a) *Investigating the interactions between individuals' time and space constraints, considering intra-household interactions. The 14th International Conference on Travel Behavior Research* (IATBR), London, UK, submitted for publication consideration of Journal of Transport Geography.

Liu, C., Susilo, Y.O. and Karlström, A. (2015b) Investigating the impacts of weather variability on individual's daily activity – travel patterns: A comparison between commuters and non-commuters in Sweden. *Transportation Research Part A* 82,47–64.

Lucas, K., Mattioli, G., Verlinghieri, E. and Guzman, A. (2016) Transport poverty and its adverse social consequences. *Proceedings of the Institution of Civil Engineers: Transport* 169, 353–365.

Merz, J., Böhm, P. and Burgert, D. (2009) Timing and fragmentation of daily working hours arrangements and income inequality – an earnings treatment effects approach with German time use diary data. *International Journal of Time Use Research* 6 (2), 200–239. dx.doi.org/10.13085/eIJTUR.6.2.200–239.

Michelson, W. (2009) Variations in the rational use of time – the travel pulse of commutes between home and job. *International Journal of Time Use Research* 6 (2), 269–285. dx.doi.org/10.13085/eIJTUR.6.2.269–285.

Musselwhite, C.B.A., Holland, C. and Walker, I. (2015) The role of transport and mobility in the health of older people. *Journal of Transport and Health* 2 (1), 1–4.

Naess, P. (2006) *Urban Structure Matters: Residential Location, Car Dependence and Travel Behaviour*. The RTPI Library Series No. 13. London: Routledge.

Neutens, T., Delafontaine, M., Scott, D.M. and De Maeyer, P. (2012) An analysis of day-to-day variations in individual space-time accessibility. *Journal of Transport Geography* 23, 81–91.

Newman, P. and Kenworthy, J. (1999) Costs of automobile dependence: Global survey of cities. *Transportation Research Record* 1670, 17–26.

Niemi, I. (2009) Sharing of tasks and lifestyle among aged couples. *International Journal of Time Use Research* 6 (2), 286–305. dx.doi.org/10.13085/eIJTUR.6.2.286–305.

Papastefanou, G. and Gruhler, J. (2014) Social status differentiation of leisure activities variation over the weekend – approaching the voraciousness thesis by a sequence complexity measure. *International Journal of Time Use Research* 11 (1), 1–12. dx.doi.org/10.13085/eIJTUR.11.1.1–12.

Schafer, A. and Victor, D.G. (2000) The future mobility of the world population. *Transportation Research Part A* 34, 171–205.

Schwanen, T. (2006) On 'arriving on time', but what is 'on time'? *Geoforum* 37, 882–894.

Schwanen, T., Kwan, M.P. and Ren, F. (2008) How fixed is fixed? Gendered rigidity of space-time constraints and geographies of everyday activities. *Geoforum* 39, 2109–2121.

Susilo, Y.O. (2011) Paratransit. In: Kenneth, B., Peter, N. and Henry, V. (eds.), *A Dictionary of Transport Analysis*, 294–296. Edward Elgar Publishing Ltd, UK.

Susilo, Y.O. and Avineri, A. (2014) The impacts of household structure to the day-to-day variability of individual and household stochastic travel time budget. *Journal of Advanced Transportation* 48, 454–470.

Susilo, Y.O. and Axhausen, K.W. (2014) Stability in individual daily activity-travel-location patterns: A study using the Herfindahl-Hirschman Index. *Transportation* 41, 995–1011.

Susilo, Y.O. and Dijst, M. (2009) How far is too far? Travel time ratios for activity participations in the Netherlands. *Transportation Research Record* 2134, 89–98.

Susilo, Y.O. and Dijst, M. (2010) Behavioural decisions of travel-time ratio for work, maintenance and leisure activities in the Netherlands. *Journal of Transportation Planning and Technology* 33, 19–34.

Susilo, Y.O. and Joewono, T.B. (2017) Indonesia. In: Dominic, S. and Dorina, P. (eds.), *The Urban Transport Crisis in Emerging Economies*, 107–126. The Urban Book Series. New York: Springer.

Susilo, Y.O., Joewono, T.B., and Vandebona, U. (2015) Investigating the reasons underlie repetitive traffic violations behaviours among motorcyclists. *Journal of Accident Analysis and Prevention* 75, 272–284.

Susilo, Y.O. and Kitamura, R. (2008) Structural changes in commuters' daily travel: The case of auto and transit commuters in the Osaka metropolitan area of Japan, 1980 through 2000. *Transportation Research Part A* 42, 95–115.

Susilo, Y.O. and Liu, C. (2017) Examining the relationships between individual's time use and activity participations with their health indicators. *European Transport Research Review Journal*, 9, 26. doi: 10.1007/s12544-017-0243-y.

Susilo, Y O., Lyons, G., Jain, J. and Atkins, S. (2012) Great Britain rail passengers' time use and journey satisfaction: 2010 findings with multivariate analysis. *Transportation Research Record* 2323, 99–109.

Tacken, M. (1998) Mobility of the elderly in time and space in the Netherlands: An analysis of the Dutch national travel survey. *Transportation* 25, 379–393.

Tarigan, A.K., Susilo, Y.O. and Joewono, T.B. (2014) Segmentation of paratransit users based on service quality and travel behaviour in Bandung, Indonesia. *Journal of Transportation Planning and Technology* 37, 200–218.

Titheridge, H., Christie, N., Mackett, R., Oviedo Hernandez, D. and Ye, R. (2014) *Transport and Poverty*. A review of the evidence. London, UK: UCL. Available at: www.ucl.ac.uk/transport-institute/pdfs/transport-poverty (accessed December 2016).

Ureta, S. (2008a) To move or not to move? Social exclusion, accessibility and daily mobility among the low-income population in Santiago, Chile. *Mobilities* 3 (2), 269–289.

Ureta, S. (2008b) Mobilising poverty? Mobile phone use and everyday spatial mobility among low-income families in Santiago, Chile. *The Information Society* 24 (2), 83–92.

Zahavi, Y. and Ryan, J.M. (1980) Stability of travel components over time. *Transportation Research Record* 750, 19–26.

Zahavi, Y. and Talvitie, A. (1980) Regularities in travel time and money expenditures. *Transportation Research Record* 750, 13–19.

Zhang, J., Kuwano, M., Lee, B. and Fujiwara, A. (2009) Modeling household discrete choice behavior incorporating heterogeneous group decision-making mechanisms. *Transportation Research Part B* 43, 230–250.

7 Moving beyond informality? Theory and reality of public transport in urban Africa

Dirk Heinrichs, Daniel Ehebrecht and Barbara Lenz

A significant part of the provision and regulation of transport services in urban areas in Africa is commonly referred to as 'informal' or 'paratransit'. Informal transport operation in particular is commonly conceptualized as something that occurs without involvement of the state or government, both with respect to the market and the practices of operation and organization. This chapter raises the question whether this clear dichotomy that separates the 'formal' from the 'informal' is an appropriate guide to analyse and make sense of current practices in urban transport provision in African cities and explores alternative concepts such as institutional or urban bricolage. The chapter uses case study research to investigate how theory meets reality. It explores the public road passenger transport system in the city of Dar es Salaam that largely consists of bus services provided by private operators of minibuses (*daladala*), two- and three-wheeled motorcycle-taxis (locally known as *bodaboda* and *bajaj*) and, most recently, a bus rapid transit service. This analysis is to a large extent based on preliminary results of field research activities in Dar es Salaam, which include participant observation and conversations with stakeholders. A literature review and a document analysis of laws and regulations concerning public transport in Tanzania complement the investigation. The analysis shows that it is indeed virtually impossible to draw a clear distinction between formal and informal spheres. Instead, the study uncovers the provision and regulation of seemingly 'informal' transport services as 'inextricably related practices' of diverse governmental and non-governmental actors that are constantly renegotiated and governed by formal and informal institutions (i.e. rules). The chapter concludes that institutional or urban bricolage as a concept and 'co-creation' of rules and institutions as observable practice appear to offer a more realistic lens to understand what is usually termed 'informality'.

Introduction

The formal-informal divide: a useful conceptualization?

A significant part of the provision and regulation of transport services in urban areas in Africa is commonly referred to as 'informal' or 'paratransit'. The existence of the two different words already indicates that definitions vary, although they are sometimes used interchangeably (cp. Cervero and Golub, 2007; Diaz Olvera et al., 2012). The term 'paratransit' was originally introduced in the United States in the 1960s to refer to 'unscheduled services that complement mass public transport systems' (Behrens et al., 2016: 1). With regard to public transport in the 'Global South' most authors agree that paratransit represents a privately developed service profiting from relaxed or non-existing regulatory frameworks (Salazar Ferro and Behrens, 2015). That means that paratransit also includes services that operate with permits issued by state authorities while their operations remain widely outside official regulatory frameworks (Salazar Ferro et al., 2012).

Some authors also attribute aspects such as flexibility of the operations, small and old vehicles, lack of fixed schedules and fragmented ownership to these services as further defining elements (Salazar Ferro et al., 2012; Salazar Ferro and Behrens, 2015). Xavier Godard characterizes the paratransit sector as 'mainly self-organized' (Godard, 2013, 97), referring to case studies in several cities in West and North Africa. Kumar and Barrett (2008), who investigated urban transport systems in various cities in sub-Saharan Africa, conceive the self-regulation of public transport through operator associations 'as an industry response to the vacuum left by the failure of government to regulate the sector' (Kumar and Barrett, 2008: 17). Woolf and Joubert speak about South Africa's minibus industry as part of a second economy, which exists in parallel to the formal first one and is 'most often [regulated] by community norms or self-appointed regulators' (2013: 289).

The term 'informal transport', on the other hand, refers to 'services [. . .] operating without official endorsement' (Cervero and Golub, 2007: 446). Following Portes et al. (1989) these services are arguably legal in nature, but are carried out without deferring to the state regulatory system. Informal transport operation is commonly conceptualized as something that occurs without involvement of the state or government, both with respect to market entrance (allocation of new routes, assignment of territory) and the practices of operation and organization (management of terminals, timetables, fares, dealing with accident situations, etc.).

Other authors contest this perception that the 'informal sphere' somehow exists and functions outside a 'formal sphere' for being negligent of local cultural realities (Roy and AlSayyad, 2004). They point out that formal regulation is never completely absent. In the field of public transport, for example, some kind of formal technical requirements (e.g. vehicle inspection) or fare regulations do exist (Finn et al., 2011; Salazar Ferro, 2014). Finn et al. (2011) for example, consider Accra's minibuses as self-regulated but show that public actors actually actively influence the system by negotiating the fares with the sector's representatives. Although not in the African context, investigating the informal collective taxis in Buenos Aires' outskirts, the Argentinian geographers Susana Kralich and Andrea Gutiérrez conclude that the absence of official regulation of informal transport cannot be understood as the complete absence of state control. Some disciplinary action is still exercised in most cases by state actors or representatives of the state such as the police (Kralich and Gutiérrez, 2007), for example, through displacement of transport operators, closing down of stops and stands/parking areas,[1] but also through *partial* acceptance in public space. Roy as well as Yiftachel suggest that in fact the state is a crucial actor regarding the production of relations that are being considered as informal (Roy, 2005, 2009; Yiftachel, 2009). As Roy puts it: '[t]he planning and legal apparatus of the state has the power to determine when to enact this suspension, to determine what is informal and what is not, and to determine which forms of informality will thrive and which will disappear' (Roy, 2005: 150). The state, however, does not only appear as disciplinary institution but also as 'a locus for the negotiation and legitimization of spatial claims' (Anjaria, 2011: 64). In this way, the state does not change the official regulations but still actively negotiates the conditions for informal economic activities. To describe this practice John Cross introduces the term 'semiformality' (Cross, 1998: 35).

Following these arguments, we suggest to move away from the clear dichotomy separating the 'formal' from the 'informal' and, instead, to conceptualize and to define the provision and regulation of seemingly informal transport services as 'inextricably related practices' of diverse actors that are constantly renegotiated and regulated by formal and informal institutions[2] based on factors such as individual resources, power relations, agency and legitimacy (Etzold et al., 2009; Etzold et al., 2012; Herrle and Fokdal, 2011).

Moving beyond the informal-formal divide: regulation of transport as 'institutional bricolage'

Thinking about and conceptualizing ways to manage and regulate services or the use of resources is a focus of institutional theory and policy. Single or collective actors (or organizations) interact with one another, according to and bounded by a set of institutions or rules (North, 1990). These institutions shape the individual perception of situations/interactions, in which actors take part, and enable or constrain their respective actions (Douglas, 1987). The theoretical perspective of New Institutional Economics, in particular, has been influential in designing and defining principles for policy approaches and their emphasis on formal public structures with clear boundaries, transparency and the codification of rules through written contracts, legal arrangements and the specification of rights as well as the actors. These principles have become mainstream for developing and implementing projects worldwide. One may argue that precisely because of its focus on 'formal' institutions, institutional economics has contributed to separate the 'formal' from the 'informal'.

This conceptualization has been challenged for different reasons. In her analysis of natural resources management in rural communities in Africa, Frances Cleaver observed that state-induced formal regulations were not applied as envisaged but rather mixed with existing forms of community management. Cleaver raises a set of arguments (Cleaver, 2000, 2001, 2002, 2007). In her view, such external policy designs are often (too) static concepts of local and/or indigenous traditions and social relations. In being so, they shed little light on processes of transformation and evolution. Also, they simplify and reduce often complex realities and diversities of interests, with the effect that they may end up favoring the interests of particular groups or actors while overlooking and neglecting others.

Building on earlier work (Bourdieu, 1977; Giddens, 1984; Douglas, 1973, 1987), Cleaver seeks to illustrate how social theory of collective action can enhance and go beyond 'rational choice' approaches. She puts forward the term 'bricolage' to describe such complex and constantly evolving arrangements that govern (in-) formal practices (Cleaver, 2007; Cleaver et al., 2013). Her work shows that '[p] eople's relations – in their economic, political and social dimensions – are negotiated not just through formal institutions, but through households, networks, cultural norms and practices, through conflict, trust and cooperation, modes of power and authority, and exclusion, and through relations of gender, age and religion' (McFarlane, 2012: 103). Bricolage can roughly be understood as the interplay of socially embedded or informal rules and modes of behavior as well as legal rules and rights, or 'dynamic hybrids of the modern and traditional, the formal and informal' (Cleaver et al., 2013: 5). This also includes the transfer, borrowing and adoption of institutions and traditions that have originally been created in other contexts but that show structural similarities and 'fit' to the context in question. Cleaver terms this 'leakage of meaning' based on Douglas' notion of 'institutional leakage' (Cleaver, 2001: 29; Douglas, 1973: 13). The resulting arrangements are an outcome of processes of actively combining, reassembling and also redesigning existing rules, traditions, norms and values but also legal rights, contracts or formal sanctions in order to create new governance structures which are responsive to changing environments (Cleaver et al., 2013: 5; Etzold et al., 2012: 188; McFarlane, 2012: 103). The concept of bricolage, therefore, does not assume rigid institutional or regulatory frameworks but rather points to a creative process of constant negotiation that continuously challenges existing arrangements. The capacity of a bricoleur – who can be a state actor or a private individual or group – to shape such configurations does vary widely and depends highly on existing power relations and the bricoleur's knowledge, agency and resources, resulting in 'a rich diversity of pliable institutional arrangements' (Cleaver, 2001: 29).

The key characteristics of bricolage can be summarized as follows:

- The institutions that shape and constrain action can be characterized rather as social relations than formalized codified rules.
- However, bricoleurs constantly negotiate and adapt traditional arrangements and integrate them with new 'formal' institutions resulting from state regulation; thus, institutions are not static and intersect formal and informal, traditional and modern, or better bureaucratic and social domains (Cleaver, 2002: 13).
- Different bricoleurs are likely to apply their knowledge, power and agency in different ways; hence the result is a rich diversity of institutional arrangements.

Objective and methodology

The subsequent sections of this chapter adopt this perspective of 'institutional bricolage' for exploring the public transport system in Dar es Salaam and to illustrate the dynamic character and the complex and fluid process of institutional co-creation. Using the concept of bricolage and its key characteristics, this chapter also seeks

Table 7.1 Data sources, methods and insights of the explorative study in Dar es Salaam in 2015

Data Source	*Method*	*Insights into . . .*
Conversations with motorcycle-taxi/*bodaboda* and *bajaj* drivers and associations	Explorative talk	Self-regulation, capital-labour relations, (non-) compliance with rules and regulations
Conversations with vehicle fleet owners	Explorative talk	Capital-labour relations, (non-) compliance with rules and regulations
Conversations with public officials	Explorative talk	State regulation, (non-) compliance with rules and regulations, road safety issues
Conversation with researcher	Explorative talk	Public transport sector, challenges, conflicts
Conversation with NGO	Explorative talk	Road safety issues, challenges and conflicts
Exploration of minibus/*daladala* service	Participant observation	Self-regulation, (non-) compliance with rules and regulations
Exploration of motorcycle-taxi/*bodaboda* and *bajaj* service	Participant observation	Self-regulation, (non-) compliance with rules and regulations
Laws and regulations	Document analysis	State regulation
Academic literature	Literature review	State and self-regulation of minibus/*daladala* services

Source: Own design

to ground the 'theory' in practice. The analysis builds on empirical insights from Dar es Salaam gathered during a pre-study in the fall of 2015 as part of a research project on 'informal transport' in urban Africa.[3] The qualitative methodology and the analysis are based on a grounded theory approach and the development of a coding scheme (e.g. Cope, 2009; Tan, 2010). Field research methods encompass participant observation of public transport services and explorative talks with more than 30 motorcycle-taxi ('moto-taxi') drivers, chairmen of nine different moto-taxi drivers' associations, two moto-taxi vehicle fleet owners, and discussions with two city officials, a local transport researcher and a representative of a local NGO. A literature review and a document analysis of laws and regulations concerning public transport in Tanzania complement the investigation (Table 7.1).

Public transport in Dar es Salaam

The city of Dar es Salaam, currently home to approximately 4.5 million urban dwellers, is the economic and political centre of Tanzania. Because of fast economic and population growth accompanied by social change and a growing middle class, mobility needs have increased and altered intensely. Therefore, the availability and expansion of transport means and infrastructure – also in light of continuing urban sprawl along the major arterial roads into the peri-urban areas – are a main focus of the discourse on urban transformation. Studies on the public transport system of Dar es Salaam have emphasized the challenges that are associated with the changes indicated previously: increasing individual car ownership, insufficient road infrastructure, chronic traffic congestions and the question of social accessibility to and exclusion from mobility options. The incidence of frequent traffic accidents, in many cases but not exclusively linked to the widespread use of motorcycles, is another key issue (cp. Diaz Olvera et al., 2003; Kiunsi, 2013; Lizárraga et al., 2014; Melbye et al., 2015). In response to these challenges, public authorities have recently introduced new strategies by expanding, transforming and regulating the public transport system. This involves the introduction of a bus rapid transit scheme, including formal integration of private minibuses – so far the 'backbone' of public transport in Dar es Salaam (Schalekamp et al., 2008) – and steps towards the regularization and formalization of the growing motorcycle-taxi sector.

These interventions do not come without modification of the existing rules/institutions and the governance of the public transport system, as will be shown in the following discussion. Firstly, the public transport modes as well as drivers of change, that is, public and private actors, are briefly introduced. Secondly, formal regulation of services, self-regulation and the emergence of institutional bricolage are traced using the minibus- and motorcycle-taxi sectors as examples.

Overview of transport modes

The public road passenger transport system largely consists of bus services provided by private operators of minibuses (*daladala*) and to a much smaller degree by the parastatal *Shirika la Usafiri Dar es Salaam* (UDA). The rise of private minibuses began in the early 1980s when state-funded public transport declined in the context of economic crises and structural adjustment policies. At that time UDA had the monopoly on bus services in the city but for several reasons could not cover the demand, which grew rapidly as a result of strong population and spatial growth. Over time, private operators were allowed to enter the market, however, in the beginning only as sub-contractors of UDA. The private buses, mainly imported Japanese brands with capacities of approximately 18 to 35 seats (Kumar and Barrett, 2008: 67), became crucial and their numbers increased almost steadily, reaching several hundred in the early 1990s and more than 3,000 at the end of the decade (Kumar and Barrett, 2008; Rizzo, 2002). Current numbers are difficult to define but the rise of minibuses has continued up to now with estimates of 6,000 to 7,800 *daladala* in operation (cp. Ka'bange et al., 2014; Mrema, 2011). Simultaneously, the number of public buses operated by UDA decreased from more than 250 in the mid-1970s to less than 20 at the end of the 1990s (Rizzo, 2002: 135). Privatization since 2011 and investment by the private Tanzanian *Simon Group Limited* have led to an increase in the number of buses operated by UDA to about 400 in 2014. In addition, the company has recently declared its intention to invest in a new fleet of 3,000 buses to operate within the city and beyond, which may have a strong impact on the minibus sector (Rizzo, 2015: 268).

The minibus sector accounts for estimated modal split shares of around 40% of all passenger trips, based on somewhat outdated data from the early 2000s (Ka'bange et al., 2014: 177). The buses cover large parts of the city, particularly the arterial roads connecting the city centre with the suburban and peri-urban areas as well as the trunk roads within the inner urban areas.

The existing bus system is currently supplemented and partially replaced by a bus rapid transit (BRT) scheme, termed *Dar es Salaam Rapid Transit* (DART), and carried out as a public-private partnership model. It will be established within six phases to cover major trunk and arterial roads of the city in the coming years. Constructions of phase one, that is, bus lanes, stations, terminals and related infrastructure, have been completed and operations began in May 2016 with the introduction of high-capacity buses that offer interim services (ITDP, 2016; Ka'bange et al., 2014; Mzee and Demzee, 2012). Once the BRT system is fully operational it is assumed that it takes over customers from the minibuses on the routes where it is introduced, while the minibuses will be excluded from these routes and are supposed to deliver trunk-feeder services to the BRT stations on separate routes. These plans already had and will have impacts on the minibus sector. According to estimates, up to 3,100 *daladala* will be replaced in the context of DART and many transport workers (and bus owners) will lose their sources of income (Ka'bange et al., 2014; Rizzo, 2015).

Apart from the bus sector and in addition to motorcar-taxi services, motorcycle-taxis – that is, motorized two- and three-wheelers, locally known as *bodaboda* and *bajaj* – fulfil additional functions within the public transport system. Their quantity, too, is difficult to assess as reliable data are not available. However, from existing studies on urban transport as well as own personal observations, it can be estimated that their numbers have been increasing rapidly in recent years (Kiunsi, 2013). Moreover, the *Automobile Association of Tanzania* (AAT) recorded the registration of more than 23,000 motorized three-wheelers and more than 185,000 motorized two-wheelers in their import statistics for the year 2014 alone (AAT, 2015), many of which are likely to be deployed for moto-taxi services in Dar es Salaam and elsewhere.[4]

Owing to their flexibility and adaptability, motorcycle-taxis have become eligible in the context of a worsening traffic situation as a means to circumvent congestions and to reach destinations in time. This is often accomplished by non-compliance to traffic rules, though, which is likely to contribute negatively to the widespread occurrence of road accidents.[5] Furthermore, the moto-taxis serve as feeders connecting unserved residential and commercial areas with trunk road bus services or they link destinations which are not directly covered by bus routes and involve multiple transfers otherwise.[6]

Overview of actors

Small-scale entrepreneurs make up the large share of private bus owners. They own one to four vehicles, while there are fewer owners of larger fleets. Altogether there are approximately 3,000 private bus owners (Kumar and Barrett, 2008: 67; World Bank, 2011). Since the 1980s, the owners are organized in the *Dar es Salaam Commuter Bus Owners' Association* (DARCOBOA) to lobby for their interests vis-à-vis the government and to coordinate their actions. Bus workers (i.e. drivers, conductors and assistant workers) are organized in unions and associations. However, the coordination of labour interests has never reached a level that is comparable to the bus owners' state of organization and their negotiation power. Many transport workers therefore face difficult working conditions and insecurity despite recent efforts to make formal work contracts compulsory and thereby improve working conditions (cp. Rizzo, 2002, 2011, 2013).

In light of the introduction of the BRT system, the government and the DART agency that is overseeing the system's implementation have sought a dialogue with private bus owners to prevent or ease disputes with the *daladala* sector. These operators were given opportunities to participate in the new system by forming consortia to bid for shares in the operation of the new BRT system or to provide feeder services. However, it remains unclear if and how many *daladala* operators (i.e. owners and bus workers) will be included in the new system and how many will be replaced. One obstacle among others is the fact that operators have to provide their own capital to be able to form companies in order to bid for shares in the system. There will also be no compensations for the loss of routes (Mfinanga and Madinda, 2016; Rizzo, 2015; Schalekamp et al., 2008). The newly formed UDA Rapid Transit, a conglomerate of UDA; the *daladala* owners' organization DARCOBOA; and the *Association of Transporters in Dar es Salaam* (UWADAR) has received concession for offering interim services in the first phase of the operation of the new system until full services commence (Ferdinand, 2015). UDA itself has lately seen a restructuring process as a result of the aforementioned investment of the Simon Group Limited, which now holds large shares of the company that was earlier exclusively owned and managed by the national government and the Dar es Salaam City Council (Songa, 2014).[7]

Operators of motorcycle-taxis can be classified into owners, owner-drivers, renters and hire-purchasers. Although some drivers work individually, most moto-taxi drivers are members of associations, which may vary in size and degree of organization. While individual drivers cruise the city for customers and/or work with regular customers, 'moto-taxi' groups offer their services from more or less permanent stands. Despite its relatively young age it can be noted that the sector shows rather advanced forms and levels of organization. Besides the founding of primary stand/drivers' associations, insights into the sector exposed that higher-level moto-taxi associations are in the formation process, for example, in the Kinondoni Municipality, to coordinate overall issues of moto-taxi workers and drivers' groups. On the other hand, owners (of larger numbers) of vehicles are sometimes loosely allied in groups, too, in order to coordinate business strategies.[8]

Government institutions and administrative actors, respectively, have started to officially recognize moto-taxi services as legitimate means of public transport and have taken measures to control it. These regulatory authorities, which are also accountable for bus sector regulation and its enforcement, encompass a number of actors. The Surface and Marine Transport Regulatory Authority (SUMATRA) sets rules for transport regulation and is subordinated to the Ministry of Transport. Local government institutions (municipalities in this case) approve applications and hand out road licences for commercial passenger transport businesses on behalf of SUMATRA. The Tanzania Revenue Authority issues motor vehicle licences to vehicle owners, while the traffic police approve roadworthiness and enforce traffic rules. How different stakeholders constitute and govern the public transport sector is expanded on in Table 7.2.[9]

State regulation, self-regulation and institutional bricolage

Both minibus and motorcycle-taxi operations have commenced as 'informal' services and, because of high demand, managed to self-integrate into the public transport system. Over time they received increasing attention by public authorities and became subject to official recognition and then regulation. However, theory and practice of formal regulation differ substantially because many operators are not reached by, or do not (fully) comply with, existing regulation. The regulatory environment itself has been subject to dynamic change and has been characterized by co-existence and amalgamation of informal and formal practices since the occurrence of the services.

At first, *daladala* operated in a setting that was characterized by a high demand for transport means and a varying degree of recognition by the government. In the context of economic crises and public transport decline, authorities came to acknowledge alternative transport means as essential. UDA issued sub-contracts to *daladala* operators in order to solve the problem of unmet demand. However, by far not all operators were registered in this way, and the government at the same time intended to maintain state monopoly for UDA. Many *daladala*, therefore, continued to operate outside of official recognition. This pattern of registered and unregistered operators persists up to now, although over the years the numbers of registered minibuses have considerably increased as the government objective for state monopoly for UDA has been abandoned (cp. Kombe et al., 2003; Rizzo, 2002).

Similar dynamics can be observed when looking at the motorcycle-taxi sector. Although these public transport services in this particular context appeared in significant numbers roughly two decades later, they too were initially provided 'informally', and public authorities came to acknowledge their contribution to meet public transport demand only at a later stage. However, high incidences of traffic accidents, non-compliance and occurrences of the use of motorcycles for criminal activities have at the same time led to discontent with this mode of transport and have to some extent resulted in restrictive measures, as will be discussed later.[10]

Looking at both services – minibuses and moto-taxis – partially co-existing trends or phases of sector governance can be identified which comprise state regulation and formalization processes as well as forms of self-regulation and self-integration of services into the public transport system. In many respects, this leads to regulatory settings in which informal and formal rules/institutions

Table 7.2 Actors of the public transport system in Dar es Salaam

Transport operators and organizations	*Status/responsibilities/assumed interests*
Minibus/*daladala* owners	Securing profit, protection of market shares
Minibus/*daladala* workers	Drivers, conductors, assistant workers; securing income
Motorcycle-taxi/*bodaboda* and *bajaj* owners	Vehicle fleet owners and owner-drivers; securing profit/income
Motorcycle-taxi/*bodaboda* and *bajaj* drivers	Renters and hire-purchasers; securing income (and gaining ownership of vehicle)
Motorcar-taxi operators	Securing profit/income
Shirika la Usafiri Dar es Salaam (UDA)	Public bus services; securing profit, increase of market shares
Simon Group Limited	Owner of UDA; securing profit, increase of market shares
UDA Rapid Transit	Concessionaire for DART interim services (including UDA, DARCOBOA and UWADAR); securing profit, increase of market shares
Motorcycle-taxi/*bodaboda* and *bajaj* drivers' associations	Coordination and internal regulation of drivers' activities, lobbying for drivers' interests
Bus owners' associations (DARCOBOA, UWADAR)	Lobbying for transport owners' interests
Transport workers' associations/unions Transport Workers Association (UWAMADAR), Tanzania Communication and Transport Workers Union (COTWUT)	Lobbying for transport workers' interests
Local and international NGOs	Initiators of drivers' trainings; reduction of traffic accidents

State actors	*Status/responsibilities/assumed interests*
Ministry of Transport	Formulation of national transport policy, improved transport supply
Surface and Marine Transport Regulatory Authority (SUMATRA)	Formulation of transport regulation, issuing of road licenses, allocation of bus routes
Tanzania Revenue Authority (TRA)	Issuing of motor vehicle licenses for and registration of vehicle owners
Municipalities	Issuing of road service licenses on behalf of SUMATRA, registration of moto-taxi stands
Traffic police	Enforcement of traffic rules and regulations, initiators of drivers' trainings, reduction of traffic accidents
National government of Tanzania	Owner of UDA; improved transport supply
DART Agency	coordinating agency; implementation of DART
Mtaa leaders/offices	In part, registration of moto-taxi-drivers' associations

Source: Own design

interlink and create versions of institutional bricolage. These stages will be explicated in the following discussion, beginning with formal aspects to allow for an understanding of the regulatory environment in which informal organizational forms emerged and keep evolving. In so doing the multifaceted set of intertwining informal and formal practices is roughly outlined.

Codified rules and formalization of services

The formal regulatory environment is characterized by a rather complex and dynamic group of state actors with changing responsibilities and an assumed lack of coordination (Lizárraga et al., 2014). The formal regulation of *daladala* services is based on the Passenger Transport Regulation 2007, which contains standards and specifications with which bus operators must comply. These include roadworthiness inspections, the application for commercial license, road service license and third-party insurance as well as a number of details for the practical implementation of passenger transport services by operators (Republic of Tanzania, 2007). Regarding commercial moto-taxis, regulations were introduced subsequently with the Transport Licensing (Motor Cycles and Tricycles) Regulations 2010, which contain comparable specifications on formal registration and licensing. For full registration, the owner and driver of the respective vehicle must present a written contract to the issuing authority, that is, the municipalities in this case. Moreover, a letter proving membership in a drivers' association and a suggestion for the intended area of operation must be provided, the latter of which the respective local authority may then approve (Republic of Tanzania, 2010).

The spatial assignment of the services results from the official registration of parking areas for moto-taxis, that is, 'designated areas' (Republic of Tanzania, 2007) and the allocation of bus routes by the city authorities. In 2009 *daladala* served approximately 126 routes (World Bank, 2011) and along these routes there exist designated bus stops as well as terminals where passengers can board and disembark. Stands of *bodaboda* and *bajaj* have initially not been earmarked but local government officials are now in the process of registering them gradually.[11] The registration of stands and the allocation of drivers might not, however, keep up with actual dynamics as still new parking areas are established and because the composition of drivers' groups and associations can change. Moreover, there exist no official (but unrecognized) stands in the city centre as the government has announced the ban of commercial motorcycles from there for security purposes and in light of the negative perception they face to some extent.[12]

Formalization of services furthermore includes the setting of tariffs as well as guidelines on equipment of drivers and crews, respectively. With regard to fares, SUMATRA has fixed amounts for routes, which in case of bus services are indicated on the exterior of the vehicles. The prices for moto-taxi services have originally been fixed as fare per trip (World Bank, 2011), but common practice is negotiation in advance of a passenger trip or the setting of fares for certain routes within the drivers' associations.[13] With regard to standards and equipment, the regulations provide guidelines for vehicles, drivers and conductors of minibuses as well as for moto-taxi drivers. In the case of *bodaboda*, for example, the drivers must be equipped with reflecting vests and helmets and provide additional helmets for customers. Crews of minibuses have to wear uniforms (Republic of Tanzania, 2007, 2010).

Owners' and drivers' associations and unions, respectively, are legally recognized as formal stakeholders representing the diverse interests of supply-side actors in the transport sector. While these organizational forms are still in a process of formation within the moto-taxi sector[14] and do not seem to have a strong influence on policies and strategic transport planning yet, they play a role in the minibus sector. Workers are to some degree organized through workers' unions and associations such as the Transport Workers Association (UWAMADAR) and the Tanzania Communication and Transport Workers Union (COTWUT), while bus owners are strongly organized via the associations DARCOBOA and the newly formed UWADAR (Ferdinand, 2015; Rizzo, 2013). Although it is not clear to what extent they can affect policy formulation, their influence can be comprehended by looking at the recognition they received in the negotiation process of the bus rapid transit scheme (e.g. Republic of Tanzania, 2014).

Major challenges exist with drivers' and also vehicle owners' compliance on the one hand and with enforcement of regulations and traffic rules on the other (see further discussion later). As SUMATRA and municipalities do not provide staff for this purpose, enforcement of regulations is carried out by the traffic police almost exclusively. However, as the traffic police itself is understaffed, it lacks the capacities to do this thoroughly (cp. World Bank, 2011).[15]

A formal measure to improve drivers' compliance with licensing procedures and traffic rules and therefore their contribution to the reduction of traffic accidents are drivers' seminars and trainings, which are largely directed at motorcycle-drivers and have already seen many participants. These measures are organized jointly by local authorities and NGOs, and in addition to educative actions and awareness campaigning, they offer practical driving training. However, it is unclear if these trainings can and will be offered on a permanent basis and to what extent they will actually lead to compliance as the reasons for non-compliance are presumably manifold.[16]

The next section lays out how these formal and codified institutions are incorporated, altered, complemented or replaced in everyday actions of local actors, leading to a dynamic 'meshwork' (McFarlane, 2012: 101) of formal/codified and informal/social rules and practices.

Forms and processes of institutional bricolage

In addition to formal regulation of services, a number of organizational aspects have been developed rather informally, based on social norms or traditions and out of practical reasoning but co-exist or merge with formal regulatory aspects.

With regard to motorcycle-taxis, these include first of all the establishing of stands, which are recognized by local authorities and widely accepted by the general public as they now make up a common feature of local public spaces and are frequently used by customers. The choice, occupation and organization of a location by a group of drivers seem to result from a rather indistinct process of negotiation, adoption and extension of existing social practices: The spatial organization of stands, which resembles that of long-established motorcar-taxi services, results from practices of queuing along trunk roads and/or the creation and arrangement of parking areas at or near bus stops and terminals, on road reserves, intersections, gateways, etc., and thus the usage of existing formal infrastructures. The stands serve as the point of origin for the moto-taxi operations and drivers, who are associated with a group; they generally return there after trips to queue for the next customer as their membership in a drivers' group or association in general obliges them to do.[17]

For the drivers this secures, despite observed incidences of over-supply, a guaranteed access to customers and it also contributes to the defence of the location vis-à-vis rival drivers as well as to the groups' legitimation as a publicly and by now in most cases formally accepted occupant of the stand. Thus, the relation of formal and informal practices is reciprocal; while drivers apply social norms such as queuing and adhering to internal rules of their group, as well as adopt formal organizational forms of founding associations and stands, state authorities recognize and register drivers' groups and their stands, which have been established informally, that is, without their prior approval.

While it seems that the formation of drivers' groups originally was to secure this access to customers in a certain location and prevent conflicts between competing drivers, by now many formerly loosely associated groups developed into associations with a higher degree of organization and additional tasks. In many cases, drivers' groups function as savings associations providing social insurance services for their members, on a rather simple level though, or even access to small loans – also a common practice among other transport workers such as motorcar-taxi drivers. This function resembles or even seems to be an adoption of the Savings and Credit Corporative Organization (SACCO) scheme, a sort of microfinance model, which has been introduced in Tanzania in the 1990s and spread ever since (Bwana and Mwakujonga, 2013).

Moreover, drivers' associations regulate behaviour between members, behaviour towards customers, conflict resolution and sometimes set tariffs for certain routes or distances. In some cases, the internal codex is documented in the form of written statutes. In other cases, internal rules are only verbally agreed upon. The scope of rules differs depending on the individual groups' level of organization. The codification of rules and the appointment of chairmen, treasurers and commissioners for the keeping of internal documents/statutes, which occurs in many cases, indicate processes of institutional development, self-regulation and formalization of services. These processes are partly stimulated, however, by public institutions, for example, by distributing blueprinted statutes during driver training sessions, which moto-taxi drivers can use as a basis for the organization of their associations.[18] The (partial) adoption of these pre-written statutes and their modification by the drivers' associations over time represent another example of institutional bricolage.

Furthermore, in many cases, drivers' associations make use of local government structures and register with so-called *mtaa* offices, that is, the lowest level of local government with basic administrative functions. Members of drivers' associations explicated that this (seemingly) non-obligatory practice serves to build up legitimation as a locally accepted association and to receive government support more easily if needed.[19] Another explanation for relying on the *mtaa* leaders may be the important political and social role they have played for a long time as spokesmen, mediators and coordinators of community activities in the local governance system (cp. Kyessi, 2005).

An additional aspect is the negotiation of work contracts or agreements between vehicle owners and drivers who intend to rent or hire-purchase a motorcycle. The agreements – which do not necessarily follow or completely apply existing labour laws (cp. Republic of Tanzania, 2004) – contain specifications on the amount the driver has to pay on a daily or weekly basis to the owner and provisions for the bearing of costs for fuel and maintenance; but in most cases that were encountered, it does not contain additional specifications, for example, for provisions on working conditions or on how cases of vehicle loss/theft or vehicle defects are settled. Similar arrangements exist in the minibus sector, although here it seems more difficult to gain sufficient revenues (cp. Rizzo, 2002, 2011). In case of moto-taxi hire-purchasers the contract also contains an agreement regarding the period of time after which the ownership is transferred from the owner to the hire-purchaser. The regularly paid amounts and the lengths of contracts again are situated in very similar ranges, only differing to a larger degree between the two modes of two- and three-wheelers because of the higher costs associated with three-wheelers. The setting of these contract details seems to be the result of negotiation processes between stakeholders in some places that was then applied in practice and subsequently must have been adopted by other operators. One owner of more than 10 moto-taxis indicated that these amounts and contract terms are at least in some cases cooperatively agreed upon by owners' groups and then incorporated into agreements with drivers.[20]

Conversations with several respondents showed that the formation of a contract – in many cases orally or, as requested by formal regulation, in written form – comes about through social kinship networks which are often very essential for securing livelihoods not only in this local context. The networks serve at least three purposes: First, to bring together those who intend to employ someone for their motorcycle and those who look for a job opportunity with low entrance barriers. Second, they can serve as warrantors and thus provide securities for both the owner and the potential driver who can rely on these networks in cases such as conflicts over costs, income loss, vehicle theft, etc. Third, when owners predominantly employ drivers to whom they are related, this can serve to strengthen economic opportunities within the extended kinship network. Although the agreements or contracts are a precondition for receiving a commercial license, it is unclear to what extent written formal contracts exist at all, to what extent they recognize formal regulations and labour laws, respectively, and what role they play for the actual work and business relationship between owners and drivers. This most likely varies from case to case and depends on the ways the bricoleurs – here primarily the vehicle owners and drivers – deploy their individual negotiation power, resources and agency.[21]

Similar to the above-mentioned, another important aspect of moto-taxi services in the city is occasional, sometimes regular non-compliance with transport regulations and traffic rules – especially with regard to the *bodaboda* – and the creation of extra-legal and irregular arrangements. These contain non-conformity with licensing procedures and vehicle and safety standards, defiance of traffic rules, and to some extent negotiations and bribery between members of the traffic police and those drivers who seek to circumvent rules and regulations, reduce penalties when caught or get access to inner-city areas where commercial motorcycles are only allowed under very restricted circumstances, that is, having a special license, carrying government officials or being physically disabled.[22]

The city centre ban has implications for the spatial and practical organization of moto-taxi services and the negotiation of rules, which are worth noting. It provokes individual strategies of drivers who, for instance, seek to adapt their service hours to the presence and non-presence of traffic controls or who develop strategies for picking up customers while being 'on the move' to reduce the chances of getting caught while parking the vehicle. Others use personal contacts to get informed about traffic controls or have arrangements with members of the traffic police who let them pass. While entering the inner-city area involves risks, it enables drivers to charge higher prices for trips to and from the centre. At the same time and despite the ban, the demand for moto-taxi services in this area is high, especially in peak hours. As a consequence, some moto-taxi drivers manage to integrate their service in this location by 'negotiating illegality' (Anjaria, 2011).[23]

Finally, the minibus sector, which has been officially regulated much earlier compared with the moto-taxis, is likewise characterized by an amalgamation of formal/codified and of rather informal/non-codified organizational forms, rules and practices. This, too, pertains to labour-capital arrangements, which do not necessarily follow labour laws and regulations and are subject to permanent negotiation and contestation (Rizzo, 2013), and to some extent to non-compliance with service regulations, vehicle quality and road safety standards. This includes inter alia the manipulation of fares, non-compliance with route assignations and deviation from traffic rules. It has been observed that minibuses sometimes charge higher prices than those that have been set by SUMATRA or full prices are charged for a complete trip, although the route is not served until the final destination and passengers are forced to disembark before, while the crew returns for the next trip. Also, minibus conductors often refuse pupils or elderly persons access to the service as these customer groups are granted fare discounts by the government. The reason for this practice may be that bus crews otherwise face revenue losses, which can become critical because they are under constant pressure to pay the owner and still keep an income for themselves at the end of the day (Ka'bange et al., 2014; Lizárraga et al., 2014). Limited negotiation power of minibus workers within capital-labour relations or vis-à-vis government institutions thus leads to a situation in which at least those formal and/or social rules, which to a certain degree can be negotiated, controlled or at least influenced by them, become subject to alteration and/or non-recognition.

The same applies to the practices of changing routes or the serving of other routes, which have not been allocated to the respective bus crew. In many cases, minibuses operate even completely without registration and license and compete with formally registered ones (Ka'bange et al., 2014).

Deviation from traffic rules, such as speeding and reckless driving to quickly reach bus stops and to be able to start the next tour in time, constitutes another widespread and long-established feature of rivalry between minibus crews who compete for customers. This is possible as there are no timetables and buses start their trips at their respective starting points when filled up with customers, stop often irregularly along the way, and start the return tour from the destination according to the same principle (Lizárraga et al., 2014).[24] Rizzo (2011) showed that an association of minibus workers in the early 2000s, in an attempt to self-regulate services, tried to resolve fierce competition and conflicts between crews by organizing services with the help of a queuing principle and the sharing of benefits and administrative responsibilities at bus stops, such as collecting member fees and counting services trips, but failed to permanently sustain this formal organizational level.

Conclusions

With a perspective of 'institutional bricolage' this chapter explores the public transport system in Dar es Salaam. The analysis of the regulation, organization and practices of public transport modes in Dar es Salaam indicates how unrealistic it is to draw a clear distinction between formal and informal spheres as observers of the 'informal sector' often do. Such a dichotomy does not reflect adequately the various and often intangible forms of co-existence, merging and application of state regulation and of (implicit) rules and practices which stem from established and changing social norms, experiences or traditions. In the case presented, different forms of formal regulation and control of minibus and moto-taxi services seem not to cover completely all aspects of the services nor do they reach all stakeholders of the sectors. At the same time, these stakeholders are bricoleurs who contribute both to the legitimation of formal rules by adhering to them as well as to their modification or replacement by creating (or reproducing) or mixing social rules in their daily practices. This pertains to the internal organization of drivers' associations and the actual performance of services, capital-labour relations, as well as to individual practices and strategies of non-compliance to regulations and traffic rules.

Institutional or urban bricolage (Cleaver, 2000, 2001; McFarlane, 2012) as a concept and 'co-creation' of rules and institutions as observable practice appear to offer a more realistic lens to understand what is usually termed 'informality'. It enables a better understanding of the way in which rules and organizational arrangements explicitly or implicitly are established, applied, reproduced and/or modified in the field of urban public transport governance in Dar es Salaam (and in other urban contexts in sub-Saharan Africa). Beyond conceptualizing daily activities of people as bricolage, the concept can therefore also serve as an analytical tool to explore the complex and fluid relationship between the formal and the informal. And it may be useful for state actors to address the challenge of '[f]inding a way in which planning can work with informality' (Watson, 2009: 2268).

Notes

1 The terms 'stand' and 'parking area' are used interchangeably throughout the chapter. They refer to physical locations that are used by motorcycle-taxi, shared taxi or motorcar-taxi operators to park their vehicles in order to wait for customers.
2 The term 'institutions' is to be understood in an institutional economic sense as 'society's "basic rules of the game" ' (Sclar and Touber 2011: 177).
3 The project investigates governance structures and practices of motorcycle-taxi services in Dar es Salaam, Tanzania, and is funded by the German Research Foundation (duration: 2015–2018).
4 Two-wheelers are predominantly imported from China, whereas three-wheelers are mostly imported from India.
5 Own field notes: explorative talk with an NGO (December 2015)
6 Own field notes: participant observation, explorative talks with a local researcher and with an NGO (November/December 2015)
7 Because the Dar es Salaam City Council sold its shares in early 2016, UDA is now owned by the Simon Group Limited, which holds 51% of the shares; the national government holds the remaining 49% of the shares (Tanzania Daily News 2016).
8 Own field notes: explorative talks with motorcycle-taxi drivers and associations, and vehicle fleet owners (November/December 2015)
9 Own field notes: explorative talk with public official (December 2015)
10 Own field notes: explorative talks with a local researcher and with an NGO (November/December 2015)
11 Own field notes: explorative talk with a public official (December 2015)
12 Own field notes: participant observation, explorative talks with public officials and motorcycle-taxi drivers and associations (November/December 2015)
13 Own field notes: explorative talks with motorcycle-taxi drivers and associations (November/December 2015)
14 Own field notes: explorative talks with motorcycle-taxi drivers and associations (November/December 2015)
15 Own field notes: explorative talks with an NGO and a public official (December 2015)
16 Own field notes: explorative talks with an NGO, motorcycle-taxi drivers and associations (December 2015)
17 Here and in the following paragraphs: own field notes: participant observation, explorative talks with a public official and motorcycle-taxi drivers and associations (November/December 2015)
18 Own field notes: explorative talks with motorcycle-taxi drivers and associations (November/December 2015)
19 Own field notes: explorative talks with motorcycle-taxi drivers and associations (November/December 2015)
20 Own field notes: explorative talks with motorcycle-taxi drivers and associations and vehicle fleet owners (November/December 2015)
21 Own field notes: explorative talks with motorcycle-taxi drivers and associations and vehicle fleet owners (November/December 2015)
22 Own field notes: participant observation, explorative talks with public officials, motorcycle-taxi drivers and associations and vehicle fleet owners (November/December 2015)

23 Own field notes: explorative talks with motorcycle-taxi drivers and a public official (November/December 2015)
24 Own field notes: participant observation of minibus services (November/December 2015)

References

Anjaria, J.S. 2011 Ordinary states: Everyday corruption and the politics of space in Mumbai. *American Ethnologist* 38 (1), S.58–72.

Automobile Association of Tanzania (AAT) (2015) *On the move. Official Magazine of the Automobile Association of Tanzania,* 2014–2015 (7th Edition). Dar es Salaam: AAT

Behrens, R., McCormick, D. and Mfinanga, D. (2016): An introduction to paratransit in sub-Saharan African cities. In: Behrens, R., McCormick, D. and Mfinanga, D. (eds.), *Paratransit in African Cities*, 1–25. Operations, Regulation, and Reform. London and New York, NY: Routledge

Bourdieu, P. (1977) *Outline of a Theory of Practice*. Cambridge: Cambridge University Press.

Bwana, K.M. and Mwakujonga, J. (2013) Issues in SACCOS development in Kenya and Tanzania: The historical and development perspectives. *Developing Country Studies* 3 (5), 114–121.

Cervero, R. and Golub, A. (2007) Informal transport: A global perspective. *Transport Policy* 14 (6), 445–457.

Cleaver, F. (2000) Moral ecological rationality, institutions and the management of common property resources. *Development and Change* 31 (2), 361–383.

Cleaver, F. (2001) Institutional Bricolage, conflict and cooperation in Usangu, Tanzania. *IDS Bulletin* 32 (4), 26–35.

Cleaver, F. (2002) Reinventing institutions: Bricolage and the social embeddedness of natural resource management. *The European Journal of Development Research* 14 (2), 11–30.

Cleaver, F. (2007) Understanding agency in collective action. *Journal of Human Development* 8 (2), 223–244.

Cleaver, F., Franks, T., Maganga, F. and Hall, K. (2013) *Beyond negotiation? Real Governance, Hybrid Institutions and Pastoralism in the Usangu Plains, Tanzania.* Environment, Politics and Development Working Paper Series, 61. London: Department of Geography, King's College.

Cope, M. (2009) Transcripts (Coding and Analysis). In: Kitchin, R. and Thrift, N. (eds.), *International Encyclopedia of Human Geography*, 350–354. Amsterdam: Elsevier

Cross, J.C. (1998) *Informal Politics*. Street Vendors and the State in Mexico City. Stanford: Stanford University Press.

Diaz Olvera, L., Plat, D. and Pochet, P. (2003) Transportation conditions and access to services in a context of urban sprawl and deregulation. The case of Dar es Salaam. *Transport Policy* 10 (4), 287–298.

Diaz Olvera, L., Plat, D., Pochet, P. and Maïdadi, S. (2012) Motorbike taxis in the 'Transport Crisis' of West and Central African Cities. *EchoGéo* 20, 1–15.

Douglas, M. (1973) *Rules and Meanings*. Harmondsworth: Penguin.

Douglas, M. (1987) *How Institutions Think.* London: Routledge and Kegan Paul.

Etzold, B., Bohle, H-G., Keck, M. and Zingel, W-P. (2009) Informality as agency – negotiating food security in Dhaka. *Die Erde* 140 (1), 3–24.

Etzold, B., Jülich, S., Keck, M., Sakdapolrak, P., Schmitt, T. and Zimmer, A. (2012) Doing institutions. A dialectic reading of institutions and social practices and its relevance for development geography. *Erdkunde* 66 (3), 185–195.

Ferdinand, M. (2015) *Tanzania: DART Interim Bus Services Set for Take Off.* Available at: http://allafrica.com/stories/201505110460.html (accessed 6 January 2017).

Finn, B., Kumarage, A. and Gyamera, S. (2011) *Organisational Structure, Ownership and Dynamics on Control in the Informal Local Road Passenger Transport Sector.* Available at: https://kumarage.files.wordpress. com/2015/03/2011-p-02-pr-organisational-structure-ownership-and-dynamics-on-control-in-the-infor mal-local-road-passenger-transport-sector_-proc-treadbo-12-s-africa.pdf (accessed 6 January 2017).

Giddens, A. (1984) *The Constitution of Society: Outline of the Theory of Structuration.* Cambridge: Polity Press.

Godard, X. (2013) Comparisons of urban transport sustainability: Lessons from West and North Africa. *Research in Transportation Economics* 40 (1), 96–103.

Herrle, P. and Fokdal, J. (2011) Beyond the urban informality discourse: Negotiating power, legitimacy and resources. *Geographische Zeitschrift* 99 (1), 3–15.

ITDP (2016) *Dar es Salaam's BRT Could Transform Urban Life in Tanzania.* Available at: www.itdp.org/dar-es-salaams-dart/ (accessed 5 December 2016).

Ka'bange, A., Mfinanga, D. and Hema, E. (2014) Paradoxes of establishing mass rapid transit systems in African cities – A case of Dar es Salaam Rapid Transit (DART) system, Tanzania. *Research in Transportation Economics* 48, 176–183.

Kiunsi, R.B. (2013) A review of traffic congestion in Dar es Salaam city from the physical planning perspective. *Journal of Sustainable Development* 6 (2), 94–103.

Kombe, W., Kyessi, A., Lupala, J. and Mgonja, E. (2003): *Partnerships to Improve Access and Quality of Public Transport.* A Case Study Report: Dar es Salaam, Tanzania. Leicestershire: Water, Engineering and Development Centre, Loughborough University.

Kralich, S. and Gutiérrez, A. (2007) Más allá de la informalidad en el transporte de pasajeros. Reflexiones sobre los servicios chárter en la RMBA. *Lavboratorio* 8 (20), 51–57.

Kumar, A. and Barrett, F. (2008) *Stuck in Traffic: Urban Transport in Africa (Africa Infrastructure Country Diagnostic [AICD]).* Washington, DC: World Bank.

Kyessi, A. (2005) Community-based urban water management in fringe neighbourhoods: The case of Dar es Salaam, Tanzania. *Habitat International* 29 (1), 1–25.

Lizárraga, C., López-Castellano, F. and Manzanera-Ruiz, R. (2014) Advancing towards sustainable Urban mobility in Dar es Salaam (Tanzania): A Swot analysis. *Africanology* 1 (1), 40–50.

McFarlane, C. (2012) Rethinking informality: Politics, crisis, and the city. *Planning Theory and Practice* 13 (1), 89–108.

Melbye, D.C., Møller-Jensen, L., Andreasen, M.H., Kiduanga, J. and Busck, A.G. (2015) Accessibility, congestion and travel delays in Dar es Salaam – a time-distance perspective. *Habitat International* 46, 178–186.

Mfinanga, D. and Madinda, E. (2016) Public transport and daladala service improvement prospects in Dar es Salaam. In: Behrens, R., McCormick, D. and Mfinanga, D. (eds.), *Paratransit in African Cities: Operations, Regulation and Reform*, 155–173. London, New York, NY: Routledge.

Mrema, G.D. (2011) *Traffic Congestion in Tanzanian Major Cities: Causes, Impacts and Suggested Mitigations to the Problem.* Paper presented at the 26th National Conference Tanzania in Arusha/Tanzania, 1-2 December 2011.

Mzee, P.K. and Demzee, E. (2012) ITS Applications in developing countries: A case study of bus rapid transit and mobility management strategies in Dar es Salaam – Tanzania. In: Abdel-Rahim, A. (ed.), *Intelligent Transportation Systems.* Rijeka/Croatia and Shanghai/China: InTech, 41–100

North, D. (1990): *Institutions, Institutional Change and Economic Performance.* Cambridge: Cambridge University Press.

Portes, A., Castells, M. and Benton, L.A. (1989) *The Informal Economy: Studies in Advanced and Less Developed Countries*. Baltimore: John Hopkins University Press.

Republic of Tanzania (2004) *Employment and Labour Relations Act 2004*. Dar es Salaam: The Government Printer.

Republic of Tanzania (2007) *The Transport Licensing (Road Passenger Vehicles) Regulations 2007*. Available at: www.sumatra.go.tz (18th September 2017).

Republic of Tanzania (2010) *The Transport Licensing (Motor Cycles and Tricycles) Regulations 2010*. Available at: www.sumatra.go.tz (18th September 2017).

Republic of Tanzania (2014) *Implementation of Phase I of the Dar Rapid Transit System: Report on Consultations with Existing Daladala Operators and the Mitigation Measures*. Dar es Salaam: Dar Rapid Transit Agency.

Rizzo, M. (2002) Being taken for a ride: Privatisation of the Dar es Salaam transport system 1983–1998. *The Journal of Modern African Studies* 40 (1), 133–157.

Rizzo, M. (2011) 'Life is War': Informal transport workers and neoliberalism in Tanzania 1998–2009. *Development and Change* 42 (5), 1179–1206.

Rizzo, M. (2013) Informalisation and the end of trade unionism as we knew it? Dissenting remarks from a Tanzanian case study. *Review of African Political Economy* 40 (136), 290–308.

Rizzo, M. (2015) The political economy of an urban megaproject: The bus rapid transit project in Tanzania. *African Affairs* 114 (455), 249–270.

Roy, A. (2005) Urban informality: Toward an epistemology of planning. *Journal of the American Planning Association* 71 (2), 147–158.

Roy, A. (2009) Why India cannot plan its cities: Informality, insurgence, and the idiom of urbanization. *Planning Theory* 8 (1), 76–87.

Roy, A. and AlSayyad, N. (2004): *Urban Informality: Transnational Perspectives from the Middle East, Latin America, and South Asia*. Lanham: Lexington Books.

Salazar Ferro, P. (2014) *Paratransit: A Key Element in a Dual System*. Available at: http://www.codatu.org/bibliotheque/doc/a-traduire-en-en_us-le-transport-collectif-artisanal-une-composante-essentielle-dans-un-systeme-dual/ (18th September 2017).

Salazar Ferro, P. and Behrens, R. (2015) From direct to trunk-and-feeder public transport services in the urban South: Territorial implications. *Journal of Transport and Land Use* 8 (1), 123–136.

Salazar Ferro, P., Behrens, R. and Golub, A. (2012) *Planned and Paratransit Service Integration through Trunk and Feeder Arrangements: An International Review*. Paper presented at the Southern African Transport Conference (SATC 2012), Pretoria.

Schalekamp, H. and Behrens, R. (2009) *An International Review of Paratransit Regulation and Integration Experiences: Lessons for Public Transport System Rationalisation and Improvement in South African Cities*. Proceedings of the 28th Southern African Transport Conference, 6–9 July 2009, Pretoria.

Schalekamp, H., Mfinanga, D., Wilkinson, P. and Behrens, R. (2008) *An International Review of Paratransit Regulation and Integration Experiences: Lessons for Public Transport System Rationalisation and Improvement in African Cities*. Available at: http://repository.up.ac.za/bitstream/handle/2263/ 11968/Schalekamp_International%282009%29.pdf?sequence=1&isAllowed=y (14 January 2017).

Sclar, E. and Touber, J. (2011) Economic fall-out of failing Urban transport systems: An institutional analysis. In: Dimitriou, H.T. and Gakenheimer, R. (eds.), *Urban Transport in the Developing World: A Handbook of Policy and Practice*, 174–202. Cheltenham: Edward Elgar.

Songa, S. (2014) *Simon Group: Powerful Clique Out to Grab UDA*. Available at: www.thecitizen.co.tz/News/national/Simon-Group – Powerful-clique-out-to-grab-UDA/-/1840392/2319882/-/17ts4d/-/index.html (6 January 2017).

Tan, J. (2010) Grounded theory in practice: Issues and discussion for new qualitative researchers. *Journal of Documentation* 66 (1), 93–112.

Tanzania Daily News (2016) *Dar City Council Sells Shares in UDA*. Available at: http://allafrica.com/stories/201604300008.html (5 December 2016).

Watson, V. (2009) Seeing from the South: Refocusing urban planning on the globe's central urban issues. *Urban Studies* 46 (11), 2259–2275.

Woolf, S.E. and Joubert, J.W. (2013) A people-centred view on paratransit in South Africa. *Cities* 35, 284–293.

World Bank (2011) *Fare Collection System – Dar es Salaam*. Available at: www.ssatp.org/sites/ssatp/ files/publications/Toolkits/Fares%20Toolkit%20content/case-studies/dar-es-salaam,-tanzania.html (5 December 2016).

Yiftachel, O. (2009) Theoretical notes on 'Gray Cities': The coming of urban apartheid? *Planning Theory* 8 (1), 88–100.

8 One hundred years of movement control

Labour (im)mobility and the South African political economy

Jesse Harber

This chapter uses Regulation Theory to understand the role of mobilities and immobilites in the political economy of South Africa. Historically, forced relocations and movement controls were used to regulate labour to accumulate capital. Since the end of apartheid, while deliberate state controls on mobility have largely been abandoned, the spatial legacy of those policies remain, and have in many cases been deepened by post-apartheid policy and political economy. Understanding the regulatory role of mobility can help policymakers to direct resources not only to maximise easy and just mobility, but towards an overall more just political economy.

Introduction

The economic history of South Africa is a story of selective mobilities. For more than a century, control over the movement of labour has been central to profit levels. Starting with colonial land-grabs and labour management, and continuing with apartheid 'influx control', the movement of workers in South Africa was closely controlled first to boost supply in a labour-scarce country, then as a system of 'relative primitive accumulation'. Over time the system was reformed to control mobility on the basis of (racialised) class rather than race, until formal mobility controls were abolished, and subsequently so was formal apartheid.

However, the society that emerged – and in some form, remains – was and is not characterised by an abundance of mobility. A half-century of apartheid (and previously colonial proto-apartheid) had endowed South African cities with a spatial form, and public transport systems, that perpetuate many of the immobilities of the past – particularly for the poor. A combination of spatial path-dependencies (particularly in the cities and their peripheral settlements) and subsequent state policy perpetuates the immobility of the (predominantly African) poor long after the fall of apartheid. Thus, the racial immobilities of the apartheid era have given way not to mobility but to class immobilities. These immobilities continue to be central to the political economy of the country.

This chapter uses Regulation Theory to show the role of control over mobilities and immobilities in the political economy of South Africa. It shows that the racial immobilisation of labour began as early as the pre-apartheid colonial period if not earlier, and then was deepened into the very foundation of the apartheid labour system. Transport immobility of labour allowed wage costs to be compressed in a revenue-constrained economy, and (for a time) managed some of the conflict that the system produced. This system left an identifiable 'apartheid spatial form' on South Africa, and particularly its cities, and this in turn has significantly shaped the nature of mobility since the end of apartheid.

Mobility as social regulation

This chapter proposes that controls on 'mobility', here meaning the literal ability of people to move around South Africa, have been and remain a central regulating institution of South African economy and society. It does this by deploying Regulation Theory, a theoretical approach to political economy that emphasises the role of institutions in enabling, constraining and stabilising systems of capitalist accumulation.

The key concepts in Regulation Theory, for our purposes, are the *regime of accumulation* and the *mode of regulation*. The regime of accumulation is the observable pattern of capitalist accumulation, most especially the class structure of the society at the time in question, and shaped by 'the nature or intensity of technical change, the volume and composition of demand and workers' life style' (Boyer and Saillard, 2002: 38). However there are inherent contradictions, or 'conflictual tendencies' (Jessop, 1988: 150), in any regime of accumulation, most especially class conflict and the tendency to overproduction (Boyer, 2010: 71).

The resulting instabilities would threaten to overwhelm the regime of accumulation – and sometimes do – were it not for the mode of regulation. This is the system of interlocking institutions that together stabilise the regime of accumulation. Regulationist analyses typically focus on five 'primary' institutions: the monetary regime, the wage-labour nexus, the form of competition, the method of insertion into the international regime and the form of the state (Boyer and Saillard, 2002). However, in principle the mode of regulation includes any institutions that enable accumulation and suppress the contradictions of the regime of accumulation.

Together the regime of accumulation and the mode of regulation comprise a particular *growth regime*. The archetypical analysis of a growth regime is 'Fordism', the mid-century growth regime primarily of the United States characterised by 'wage growth calibrated to grow in line with increasing productivity . . . maintained by the continuing investment of profits in advancing technology, which increased productivity at a rate that sustained both increasing consumption and investment' (Neilson, 2011: 162). This was enabled and stabilised by a mode of regulation characterised by a 'Keynesian, corporatist and welfarist' class compromise (Neilson, 2011: 162).

Capitalism is in constant motion, and both the mode of regulation and the regime of accumulation are subject to internal destabilisation – such as by increasing capital-intensity changing the role and power of labour – and external shocks. If the institutions of the mode of regulation can regulate and stabilise accumulation in the new normal, or can be adjusted to do so, then the growth regime is likely to remain. Otherwise a crisis ensues until and unless a mode of regulation can be established to enable ongoing accumulation in a corresponding regime of accumulation.

Regulation Theory is a deep and wide body of theory, and this discussion is not supposed to be a thorough explication. However, it means to show the value of Regulation Theory for understanding both continuity – by analysing the functionality of the mode of regulation to the regime of accumulation – and change, by analysing the nature of the contradictions between and within the mode of regulation and the regime of accumulation, and the systemic elaboration that restores stability.

The historical analysis that follows, which is based on secondary literature, is in short, an argument that the location and degree of mobility of South African labour has always been central to the regime of accumulation. The central years of apartheid between 1948 and the early 1970s were characterised by a regime of accumulation dependant on cheap African labour, and this was enabled by a mode of regulation that contrived to produce an abundant supply of migrant labour in certain parts of the country, while stabilising workers in areas where higher skill levels were needed. This is what is meant here by 'selective' mobility: neither mobility nor immobility, but a differentiated system of labour mobility. Over time the instabilities of this growth regime mounted, and from the resulting crisis a mode of regulation that increasingly differentiated workers based on class, rather than race, emerged. The post-apartheid mode of regulation in turn has its own methods for controlling the mobility of labour.

Although race and class immobility were of obvious and enormous importance in this history, I use the terms 'mobility' and 'immobility' throughout to refer only to the ability of people to move through space. One aim here is to show how race and class immobility were operationalised and enforced largely through controls on the literal movement of people.

The migrant labour system and the origins of apartheid

The South African system of selective mobility had its origins in the colonial period that began with Union in 1910. From then through the Second World War (and some time after), the South African economy was dominated by the 'gold-maize alliance' (Lundahl, 1989) of mining and agriculture (Lipton, 1988). This political alliance between capitals rested on a shared dependence on foreign markets and cheap labour. A combination of the technical demands of South African gold deposits and the need to import almost all capital equipment meant that capital investment was primarily directed at increases in productive capacity rather than technological shifts: 'capital-deepening (increased capital intensity) occurred primarily as part of capital-widening (extending production capacity) . . . thus the increase in the capital-labour ratio was limited compared to the [Advanced Capitalist Countries] where implementation of new technologies generally involved scrapping of existing equipment' (Gelb, 1987: 5–6).

Also in this period, unskilled African labour was scarce throughout southern Africa (Lipton, 1988), not least because peasant agriculture remained a viable alternative: 'it was thus difficult to secure the requisite labour without simultaneously raising wages' (Lundahl, 1989: 829). This is not unlike the situation of early English capitalism as Perelman describes and explicitly links to the South African case (Perelman, 2000, 2007, respectively). All this happened in the context of a fixed gold price that until 1970 demanded that the profits of mining capital in particular relied on aggressively minimising costs.

Such was the set of constraints facing the dominant fractions of South African capital around and after the Second World War. The South African state had already set up a system of extensive segregation, including the 'homelands'. These were established by the Natives Land Act (1912) as areas of exclusively African subsistence agriculture and represented reasonably straightforward primitive accumulation, by means of which a labour force was created (along with a ready supply of land). Similarly the Natives (Urban Areas) Act (1923) that introduced a pass system and 'embodied the sentiments of the Transvaal Local Government Commission of 1922 that "the native should only be allowed to enter the urban areas, which are essentially the White man's creation, when he is willing to enter and minister to the needs of the White man, and should depart therefrom when he ceases so to minister" ' (Goodlad, 1996: 1630). The later, apartheid, production of spatial immobility of the African population was thus 'predicated on an ideological edifice that dates from the colonial era.' (Dawson, 2006: 126)

Over time, political economy shifted to one of 'relative primitive accumulation' (Perelman, 2007), that is, the production of not total but variable dependence on capitalist relations in general, and wage labour in particular. As the Chamber of Mines testified to the Witwatersrand Native Mine Wage Commission in 1944:

> It is clearly to the advantage of the mines that native labourers should be encouraged to return to their homes after the completion of the ordinary period of service. The maintenance of the system under which the mines are able to obtain unskilled labour at a rate less than ordinarily paid in industry depends upon this, for otherwise the subsidiary means of subsistence would disappear and the labourer would tend to become a permanent resident upon the Witwatersrand, with increased requirements.
>
> (quoted in Wolpe, 1995: 69)

And:

> The ability of the mines to maintain their native labour force by means of tribal natives from the reserves at rates of pay which are adequate for this migratory class of native, but inadequate in practice for the detribalised urban native, is a fundamental factor of the economy of the gold mining industry.
>
> (quoted in Walker, 1948: 22)

In other words, non-capitalist subsistence agriculture became an integral part of the capitalist economy. It cross-subsidised the wages of migrant labourers, whose employers were spared the costs of reproduction: the young, the old, the infirm and the spouses were left to support themselves from the land. This was highly gendered: 'the subsidy . . . was provided by rural African women' (Bond, 2007: 8).

Additional details of the early twentieth-century growth regime are not important here. Suffice to note that the system introduced control over the mobility of African labourers and their families – confining the former to marginal agricultural areas, and selectively allowing the former to move temporarily to where wage labour was available.

However, 'the stability of an accumulation regime or mode of regulation is always relative, always partial, and always provisional' (Jessop, 1988: 151): if successful, it exhibits accumulation, growth, and 'upheaval in the methods of production and lifestyles' (Vidal, 2001: 24), all of which produce instability. In this case the aggressive extraction of surplus labour led to underdevelopment and environmental degradation in the homelands. By the 1920s the homelands could no longer provide the surplus that the system depended on to cross-subsidise wages, and as the internal tensions of the growth regime deepened, African poverty became increasingly widespread.

In addition, as Mamdani notes: 'the problem with territorial segregation was that it rendered racial domination unstable: the more the economy developed, the more it came to depend on urbanised natives' (Mamdani, 1996: 66). This resulted in uncontrolled urbanisation and 'massive overcrowding' (Dawson, 2006: 127) as the state lost control over the mobility of its subjects. As a result, increasing numbers of Africans and whites were living in reasonably close proximity and made stark the vastly better treatment enjoyed by the urban, white 'alien minority' (Mamdani, 1996: 66). This led to a cycle of conflict, repression and resistance. In the 1930s, 171 088 African person-hours were lost because of strikes; in the 1940s, 1 684 915 African person-hours (Wolpe, 1995). By the end of the Second World War the growth regime was rapidly unravelling into crisis. In 1948, the Reunited National Party (later just the National Party, and hereafter referred to as such) came to power on a platform of *apartheid*, or 'separateness'.

Apartheid, influx control and mobility

From the very beginning, apartheid was constructed to serve the economic interests of White South Africans, and particularly white capital (Posel 1991 cited in Worden, 2000). The National Party had been divided by an internal debate in the run-up to the 1948 election, with hard-line nationalists demanding total segregation – a total ban on African mobility – and pragmatists advocating a qualified segregation based on the existing migrant labour system and designed to enlarge the supply and reduce the cost of African labour. The latter prevailed.

The apartheid 'solution' to the contradictions of the mode of regulation was to double down on selective mobility: on the one hand to increase the state's repressive power to maintain control over where Africans lived and worked, and on the other to establish urban 'townships' and rural ethnic 'homelands' to house the population so controlled (Dawson, 2006).

The Group Areas Act (1950), lying 'at the very core of urban apartheid' (Mabin, 1992: 405) segregated urban areas by race and allowed mass displacement of people by planners to their racially appropriate area. The Bantu Authorities Act created 'chiefdoms' to exercise indirect state control over reserves, and the Abolition of Passes and Coordination of Documents Act (1952) in fact extended the pass laws (first introduced in 1923) to every African. The *dompas*, as it came to be known, determined where its bearer was allowed to live, work and visit, and was one of the most direct and most hated mechanisms for the control of African mobility. Labour bureaux had already been established in 1951, also to control where Africans could seek work, and under the Natives (Urban Areas) Amendment Act (1955) Africans could not live in any town where they had not been born or had worked for 15 years (or 10 years for a single employer).

The net effect was: African workers (whose availability for employment at a low wage was essential for mining and agriculture alike) could live in certain areas in a city (Group Areas Act 1950), or return to their respective 'traditional' homelands (often nothing of the sort) to live under the authority of state-sanctioned or -created 'chiefs' (Bantu Authorities Act 1951). They could look for work only where allowed by a labour bureau, enforced through the passes that they had to carry by law (Natives [Abolition of Passes and Co-ordination of Documents] Act 1952, among others). Without work, they would have to leave town. The laws governing the mobility of Africans and the overarching policy were collectively known as 'influx control'. Influx control, as well as other more obviously economic laws such as the (older) so-called 'colour bar', which reserved certain categories of labour for white workers, ensured a supply of cheap African labour for white capital and formed the core of the apartheid mode of regulation.

The spatial form in this was characterised by economic centralisation, with relatively few centres of mining and manufacturing, and residential decentralisation, with 'dormitory' communities of labour scattered as far as the laws on influx control could send them. This was the case both at a national level – with urban centres being sites of major economic activity and employment, and formal labour reserves in rural areas – and at an urban level: cities had traditional economically dominant downtowns, and 'labour reserves' of their own in the form of townships allocated under the Group Areas Act. Africans were removed by force from areas designated for other population groups. By the 1970s more than a million people had been forcibly removed from South Africa's cities (Bester, 2001).

Although the law was used to prevent the unwanted movement of Africans, there had to be a way to move people as needed (by the state and by employers; there was little or no regard for the needs of the people so moved). Africans' long commutes and low incomes meant that 'regular, efficient and inexpensive public transport was imperative to ensure that the massive displacement of the workforce did not interrupt the smooth working of the economy' (Pirie, 1992: 173).

Long-distance commuting was, therefore, both necessary and common, and the resultant investment (which could not be covered by fares because of the length of journeys and poverty of commuters) was poured into extensive, but extremely selective, public transport infrastructure: links were made between residential and economic nodes – mostly by bus and train – but stations were poorly integrated into townships and little attention was given to even basic amenities (Pirie, 1992). The net result was a transport system that moved large numbers of people across large distances, but failed to serve their broader needs and, crucially, left them with long walks and waits as part of their routine. Apart from the direct control the state exercised through its monopoly on African mobility, the drudgery this imposed on African workers was not a bug but a feature: every minute spent riding, waiting for or walking between public transport was a minute that could not be spent organising or agitating against the state or an employer.

Apart from public transport, the cities required 'the full panoply of the modern state for its administration and maintenance. The classic tools of modernist social engineering, including urban planning, public administration, and criminal justice, were all deployed' (Dawson, 2006: 126) to maintain such an extensive system of selective mobility.

Crisis

The origins, causes and nature of the economic and political crisis that beset South Africa from the 1970s were the subject of much academic controversy at the time and for many years after. Here we avoid the morass by sticking to the effects that it had on the South African labour market.

By the late 1960s the South African economy had become considerably more capital-intensive, and required a correspondingly greater supply of skilled and semi-skilled labour. Demand for the unskilled labour that had hitherto made up the great bulk of the workforce shrunk accordingly (O'Meara, 1996). The regime of accumulation had undergone a secular shift that put it increasingly at odds with the apartheid mode of regulation. The growing contradictions in the economy manifested, among other ways, in growing militancy among African workers from about 1973, and soaring unemployment (Marais, 2010). Skilled labour shortages, low levels of investment in manufacturing and a saturated domestic market for the products of local manufacturing (largely limited to luxuries) all compounded the crisis.

The homelands, the surplus of which had been systematically extracted and ecologies degraded by the demands of the migrant labour system, simply could not support the growing unemployed population with subsistence farming and meagre local economies. The labour immobility at the heart of the South African growth regime eventually simply failed: 'the idea of blockading Africans in literal peripheries was in crisis. The reality of an exponentially growing, permanent, urbanized African population had become irreversible.' (Marais, 2010: 32).

But the increasing breakdown of influx control was not enough to stem the crisis: urban work for Africans was scarce and poorly paid, and the apartheid state ever more repressive. An IMF loan in 1982 brought structural adjustment, which led to inflation peaking at 17% because of the withdrawal of consumer subsidies, tax hikes and the like. Crucially, the extensive public transport that apartheid relied upon, characterised by long distances (and therefore high costs) for poor commuters, was a prime target – as was state-financed housing such as Group Areas townships and other dormitory settlements (McCarthy, 1992). The increasing costs of apartheid at home and military entanglements abroad led to a broad fiscal crisis.

The agricultural workforce shrank by a third between 1965 and 1988 (SAIRR, 1992: 396), as did the mining workforce between 1987 and 1995. The labour-absorption capacity of the economy virtually collapsed (Marais, 2010: 86), from 90% in the 1960s, to 22% in the 1980s and 7% by the end of the 1980s (SA Reserve Bank, 1991), leaving unemployment at more than 30% by the early 1990s. The regime of accumulation had shifted radically, leaving the mode of regulation behind. The damage to the South African labour market persists to this day.

Spatial policy in a time of crisis

Just as 'apartheid required the imposition of its own spatial order on human settlement' (Smith, 1992: 9), so did (and does) the late- and post-apartheid mode(s) of regulation.

From 1971, Bantu Affairs Administration Boards (BAABs) were responsible for the management of urban African townships, and from 1977, with the Community Council Act, elected municipal councils sat below the BAABs. By this point, influx control had effectively collapsed: 'unemployment and poverty could no longer be externalized to the homelands' (Beinart, 2001: 257). The 1979 Riekert Report on 'Manpower Utilisation' marked the government's formal acceptance of African urbanisation as fact (Hindson, 1991), although not its abandonment of influx control (Marais, 2010). The report reconciled the two by advocating managed urbanisation, whereby 'qualified' Africans could move to and live in cities, while the 'unqualified' remained subject to influx control. In effect, the result was 'to tighten, not relax, the mechanisms of influx control' (Gelb and Saul, 1981: 49). The resultant policy of 'inward industrialisation' intended to drive growth using the consumptive power of urbanising African workers, especially through the mass provision of low-income housing (Lewis, 1991).

The Riekert reforms were short lived and (in hindsight) decidedly transitional. By 1982 such controls as remained were straining in the face of mass urbanisation, and the mid-1980s saw two new major developments. First, large fractions of capital turned decidedly against racial controls on movement. This was exemplified by the Urban Foundation, set up in 1977 by Anglo American's Harry Oppenheimer and Anton Rupert (two of English capital's grandees) and largely responsible for coordinating the 'deracialized urbanization strategy' over the 1980s (Urban Foundation quoted in Bond, 2000: 225). Later, the Urban Foundation's 'Proposal for a National Housing Policy' would form the basis of post-apartheid housing policy (Isaacs, 2015).

Second, the President's Council Report of 1985 recommended the abandonment of all legislation regulating urbanisation by race. The replacement was 'orderly urbanization' laws governing, among other things, squatting and urban planning. Under the resulting policy the poorest members of the working class were directed into shanty towns (originally 'squatter camps', now 'informal settlements') on the urban periphery. These housed 7 million people across South Africa by the early 1990s (Beinart, 2001), while members of the wealthier working class and emerging middle class who could afford it were directed into township housing priced respectively for them. This newly stabilised workforce suited mining capital in particular, which had identified labour migrancy as a source of workplace violence, disruption and resistance (Lever and James 1987 cited in Freund, 1991).

As this process unfolded, new institutions were introduced to manage the urban space. These included Joint Management Centres (JMCs), whose 'oil-spot' approach to development combined security interventions (using police and the military) with standard-of-living upgrades (such as electrification) in order to forestall or suppress unrest (Beinart, 2001).

Another whole set of institutions developed around housing finance aimed primarily at the increasingly urbanised African population. From 1978, Africans had legal access to building society bonds and from 1983 various acts and amendments encouraged small housing loans. In Alexandra township, for example, residential building loans tripled in total value between 1986 and 1988 (Bond, 2000: 209). Thus, the policy goals of inward industrialisation began to be realised.

A combination of privatisation of state housing with a private development boom meant the state's involvement in mass housing provision shifted, although it did not necessarily diminish. The state's role became the vigorous repression of unrest, implementation of standard-of-living (and property-price) upgrades, and the extension of subsidies and subsidized finance to a new urban African petite bourgeoisie who in turn were expected to drive consumption and resist insurrection. 'Supplying a young black revolutionary with a housing bond through the disciplinarian private market (perhaps with some form of mortgage insurance program to spread the risk) is one way of tying her or him down to stable labour and community behaviour' (Bond, 2000: 198). This was to be the Urban Foundation's 'property-owning democracy'.

Meanwhile, the state was withdrawing from public transport. The transport subsidies required to move people from where they could legally live to where they could legally work were enormous, and growing exponentially (McCarthy, 1992), exacerbated by bus boycotts of increasing frequency. As pressure grew on the state to constrain its expenditure, it chose to privatise those services that would least affect its white constituency: parastatal bus companies and African housing stock (McCarthy, 1992).

With the (incremental, uneven and incomplete) dismantling of the racial regulatory order, apartheid itself came to be seen as instrumental rather than hegemonic; a means rather than an end (Giliomee 1982 cited in Greenberg, 1987): whether sincere or cynical, a revisionist approach emerged among Afrikaner academics and government reframing apartheid as a set of market-correcting interventions (Greenberg, 1987). Along with the privatisation of (hitherto racially regulated) social services (Morris, 1991), the mode of regulation shifted from depending on the explicit racial discipline of the state to the class discipline of the market. This was no less dependent, only less obviously so, on state power. And its ultimate results, as discussed later, included not the dismantling of immobility but its reproduction in a different form.

The post-apartheid spatial order

> The economic and social forces that emerged under apartheid did not suddenly expire with the advent of democracy. The legacy is embedded in conservative institutional and social practices that continue to have powerful effects, overriding many current policy aspirations.
>
> (Turok, 2001: 2350)

> The contradictions faced during the post-apartheid era are embedded in the spatial form created by apartheid-era politics of social engineering.
>
> (Dawson, 2006: 126)

The end of apartheid and emergence of a post-apartheid order resulted in 'new and powerful forces of fragmentation, represented by the suburbanisation of forms of economic activity' (Mabin, 1995 quoted in Turok, 2001: 2358). Beginning in the 1970s with suburban malls and office-blocks, and accelerating in the 1980s and 1990s, city centres diminished dramatically in economic importance (Todes, 2015). This decentralisation, deconcentration and drift from central business district (CBD)s (Turok, 2001) was accompanied by an increasing differentiation of areas by market segment. In this way, what were once suburbs became ' "totalised suburbs" in the sense that they now function independently of the central business district of the central city' (Crankshaw, 2008: 1692).

But as city centres diminished dramatically in economic importance, the largest cities *grew* overall in both relative share of output and employment, as the tertiary sector grew mostly at the expense of the manufacturing sector (Todes, 2015) resulting from, among other things, the withdrawal of various industrial policies. The economy therefore undertook a dual spatial shift: on a national scale it centralised in a handful of major cities, while on a local scale it decentralised into the urban periphery.

This was reinforced by changes in the nature of planning and policymaking. Sihlongonyane (2015) describes a broad shift over the 1980s from planning as a discipline focused on large, state-driven, top-down plans to one dedicated to producing frameworks to facilitate other actors, especially the private sector, to make decisions in a decentralised way. The role of planners therefore shifted to 'one of identifying and mediating between different interest groups involved in, or affected by, land development' (Sihlongonyane, 2015: 40).

Something similar is true across all components of South African spatial policy, not just town and city planners. For example, South African housing policy shifted from direct state-provision on a mass scale to a system of interlocking institutions designed to be an 'enabling environment' for private housing development. This included 'generous incentives for banks and developers [in] the late apartheid era' (Bond, 2000: 301), which only grew through the 1990s. For example, the Mortgage Indemnity Scheme (from 1994) guaranteed banks against bond non-repayment; Servcon (from 1994) 'rehabilitated' bonds under default; the National Housing Finance Corporation (from 1996) provided wholesale funding to retail mortgage-providers; and the National Urban Reconstruction and Housing Agency sources and guarantees financing for housing developers. This is not to say that the state's role in housing provision has diminished; rather that it shifted from direct provision to 'state-assisted, market-driven delivery' (Khan and Ambert, 2003: v).

The housing subsidy scheme introduced in the 1990s was massive in scale, and directly undermined such urban policy as existed at a national level (Todes, 2006). The focus on subsidies for individual households and the imperative to maximise houses built for the allocated budget led to large-scale developments of freestanding houses in (mostly) economically stagnant areas, where the land was cheapest (Cross, 2013). This was exacerbated by subsidy conditions, including a minimum size for houses, and a maximum proportion of the subsidy that could be spent on land – both intended to protect beneficiaries from substandard houses, but both (with the other factors) resulting in the large majority of state-subsidised housing being built in far-flung, economically stagnant areas. Thus, as cities were decentralising economically they were also decentralising and deconcentrating residentially, but these movements were to *different parts of the urban periphery*. The middle classes and wealthier compensated for this economic decentralisation with self-provision of private transport in the form of cars, and in any case, were rapidly leaving many inner cities for a number of reasons.

This in turn contributed to changes in the drivers of migration. By the 1990s unemployment was ubiquitous, and moving to the cities did not readily yield work. In fact, the national 'employment gradient' which one might expect to be positive towards city centres came in fact to be weakly and ambiguously negative (Cross, 2013). Migration came to be driven instead by the 'second-best substitute goals' of state services (Cross, 2001) largely in the form of state-provided housing (along with water and electricity) which would represent not a living, but perhaps nonetheless a higher standard of living for poor South Africans. Despite the scale of the housing programme, stable shelter remains scarce for poor families and 'permanent housing at the city edge can become a highly attractive, scarce and valuable good' that 'traps' migrants in job-poor areas (Cross, 2013: 250).

The mechanisms of this 'RDP trap' illustrate how the configuration of the market came to drive immobility. 'RDP houses' were introduced as part of the short-lived Reconstruction and Development Programme (RDP), which although long-defunct lives on in the quasi-official name for any state-provided freestanding house. Eligibility for an RDP house depends on a combined household income below R3 500 per month. Such houses, even in well-located areas (with regards to amenities and economic opportunity) 'fall short of market prices for alternative (non-RDP) low-income properties (as well as other properties in the locality)'

(Lemanski, 2011: 64), which means that despite the large demand for RDP houses, there is little opportunity to move to the next rung on the 'property ladder' without moving some distance away, which diminishes their attractiveness in the context of low incomes and very high transport costs. Thus 'in de Soto's terms this renders an RDP house "dead capital" . . . because although it can be sold . . . its utility as financial capital for household/individual economic progression is weak and hence its capital is lifeless within the broader property market' (Lemanski, 2011: 68).

The foregoing is not to say that RDP homes are not valuable to their owners. They are, as mentioned, 'second-best substitute goods' (Cross, 2001) that can in themselves dramatically raise a family's standard of living. They are also an important social and cultural asset, representing as they do to many people not just a foothold out of poverty and highly visible tangible material benefit of the democratic era, but definitive establishment of the right to legal urban residence of which Africans were so long deprived (Lemanski, 2011). Indeed, one might say that the inherent value of an RDP house is part of the problem: the disjuncture between its use value and its exchange value means that it is a step out of the very deepest poverty that can nonetheless prevent further material advancement.

The results of these and other factors has been cities with diminishing centres, and peripheries that are divided unevenly into areas of economic activity and areas of mass housing for the poor, in the form either of state-provided housing or massive informal settlements. It may be going too far to say, as Sihlongonyane does, that 'planning in the post-apartheid era . . . has tended to be a facet of the completion of the apartheid project' (2015: 39) but it is nonetheless true that 'the combined effects of a neoliberal economic logic with state interventions that tend to follow a spatial logic . . . reproduce fragmentation and sprawl, rather than promoting transformation in the interest of creating an integrated urban form' (2015: 55).

Mobility and immobility in the (post-)apartheid spatial form

The cumulative effect of all of these factors has resulted in what is sometimes called the 'apartheid spatial form' (which perhaps more properly should be the 'late-and-post-apartheid spatial form'): low-density, decentralised cities with multiple 'nodes' of activity. Working-class residential nodes are often poorly located in relation to economic nodes, and public transport moves people inefficiently between them, if at all.

The high apartheid public transport system focused on subsidised long-distance trains and buses running between distant dormitory settlements and areas of economic opportunity, including CBDs (as well as more conventional urban services for whites). This had enabled and reinforced dispersed, sprawling patterns of development with strict racial segregation. As the economy descended into crisis over the 1970s and 1980s, the withdrawal of subsidies from public transport resulted in deteriorating service and increased costs (financial and otherwise) for African commuters (McCarthy, 1992; Turok, 2001). Policymakers were increasingly looking to the informal sector to both deliver services and increase employment (Lewis, 1991), and the resulting regulatory and service gap allowed a private minibus taxi industry to emerge, and eventually come to dominate commuter transport (Marais, 2010; Todes, 2015).

This 'politically powerful and largely under-regulated fleet of over-crowded, poorly maintained minibus taxis operating irregular services' (Wood, 2015: 89 citing Salazar Ferro et al., 2013) combined with what remains of the subsidised public bus systems reinforce the fragmentation of the South African city. They ameliorate the high costs to individuals of a sprawling, deconcentrated city relative to the costs of a denser or more spatially integrated city: 'the subsidies given to transport sustain the city's polarisation and imbalances' (Turok, 2001: 2352). This effect is even stronger with minibus taxis, which (as private businesses) run primarily on routes with the highest demand, which by their nature are those that reflect the current distribution of dwellings and jobs.

These developments in housing, urban policy and transport coincided with greater cost and income pressures on poor (mostly African) workers: incomes declined in real terms between 1995 and 2000, for poor workers most of all (Leibbrandt et al., 2005); and there was a large shift towards casual and contract labour (Marais, 2010). As a result, even as apartheid was dismantled and formal restrictions on South Africans' mobility abolished, an impoverished African population nonetheless found itself treading the same routes from the same places of residence to the same places of work. As in other areas of the South African economy, as the state withdrew from the mode of regulation, the market flowed into the space it left.

The privatisation of public space

The growth of private security in South Africa began with increasing social instability in the 1970s, and accelerated with the rise of privatisation and outsourcing from the 1980s (Diphoorn, 2015). By the 1990s private security officers outnumbered public 'forces of law and order . . . by a factor of two to one' (Dawson, 2006 citing Brenner, 1998) and the industry had come to function as a 'complementary body to the apartheid state, in terms of logistics and manpower, but also in protecting white property and privilege' (Diphoorn, 2015: 203). Over the end of apartheid the industry grew enormously, driven on the supply side by growing unemployment and demobilising security forces (from both sides of the struggle), and on the demand side by a fearful white population whose government was losing the inclination and the ability to externalise crime to other, more African areas (Dawson, 2006; Diphoorn, 2015). Increasing numbers of security personnel were deployed to guard areas of residence, work and leisure.

The securitisation of white areas was not limited to guards. Homes came to be increasingly fortified by their owners, and before long entire neighbourhoods were blockaded off by 'closing street access, erecting electrified fences and high walls, as well as employing private security guards and CCTV' (Lemanski, 2004: 106). Many of these closures are illegal, either the result of residents' frustration with the slowness of municipalities processing their applications, or of no application having been made. Over the interceding decades, the result has been for cities to be. . . 'broken up into a series of discrete enclaves, with harrowing corridors of transit lying between' (Dawson, 2006: 133) to an extent that 'facilitates social exclusion, enhances urban segregation and disrupts urban planning and management' (Lemanski, 2004: 107).

The stated reasons for these securitised neighbourhoods, if any are given, centre around a fear of crime – and especially violent crime. Although 'such representations of criminality often serve as a coded form of racial discourse' (Dawson, 2006: 132), the security arrangements of privatised space are ostensibly neutral, with access restrictions (or spurious demands for identification) being levelled at all 'visitors'. The effect is to drastically limit the mobility of non-residents, and as these neighbourhoods skew wealthy and white, the result is that the same people whose entry to certain areas was limited by state force under apartheid are now excluded (or at least strongly discouraged) from those same areas by private force. As in-group inequality grows in South Africa and more African families join the middle classes, the racial character of this exclusion diminishes. However, the exclusion itself does not: 'the post-apartheid context is developing a virtually identical included/excluded socio-spatial system' (Lemanski, 2004: 109, 2011).

Conclusion

This chapter has argued, first, that the apartheid mode of regulation was characterised by controls on the mobility of South Africans – particularly Africans – in order to manage the costs of labour and distribution of the labour force, along with the escalating deployment of force to manage the conflict that resulted from this and other forms of oppression.

Second, over the course of the twentieth century a policy of racial immobility and spatial fragmentation left an identifiable apartheid spatial form on South African cities, and the persistent nature of the built environment has resulted in this being a significant source of path dependency.

Third, this path dependency is only partial: the transition from the apartheid to the post-apartheid mode of regulation has been characterised by a major shift in the axis of exclusion (and immobility) from race to class (albeit complicated by the still-racialised nature of class). Many South Africans remain limited in their physical mobility, no longer by racialised laws but by the combination of their low incomes and the high costs of movement imposed by post-apartheid cities. As Crankshaw writes: 'The post-Fordist spatial order of Johannesburg may be more unequal than its Fordist and apartheid past, but it is certainly less racially unequal. [. . .] In class terms, it is probable that the post-Fordist period is even more spatially polarised than the racial-Fordist period' (2008: 1707).

Fourth, the contribution to immobility of apartheid crystallised in the built environment is joined by forces that arose over the transition from apartheid to democracy: these include the rise of the minibus taxi industry and the securitisation of residential neighbourhoods. The path dependency has also been deepened by positive feedback loops inherent to the built environment: a fragmented city demands public and private transport systems that cope with fragmentation, which reduce the relative cost of further fragmentation, and so on.

Where there has certainly not been path dependency has been in the intentions of the state and of policymakers. In recent years, concerns about the apartheid spatial form have come to preoccupy at least some parts of the state – including many local governments. However, that preoccupation is far from universal, and has to compete with the moral and electoral imperatives to deliver services – particularly housing – at enormous scale with limited budgets. These imperatives are resulting in new plans for 'mega human settlements' of hundreds or thousands of units. In effect, the difficulty of delivering functional, integrated cities and a mobile population has led these policymakers to double down on the apartheid spatial form, with predictably poor consequences.

This is reinforced by the structure of the state: housing (as 'human settlements'), transport, economic development, are almost always separate line departments in national, provincial and local governments (and spatial planning yet another department in the latter). Separate bureaucracies, reporting lines and performance management structures (sometimes mandated by national regulation) make integrated planning and implementation extremely difficult even with the best will in the world, itself often lacking.

That said, the more progressive municipalities, provinces and parts of the national government are moving to break these 'silos', and experiment with different ways to plan and deliver their interventions. There is increasing recognition that the end of apartheid movement controls was necessary but not sufficient to achieve the free and equal mobility of South Africans, and that persistent labour immobility – including from far-flung RDP settlements – is a constraining factor in (particularly) urban economies.

However, the apartheid experience also shows that accumulation and production did not depend on ever-greater labour mobility – precisely the opposite. Nor was labour immobility an economic irrationality. Rather, variegated patterns of mobility and immobility were both constraining and enabling of the economy: favouring certain fractions of capital and disfavouring others (and overwhelmingly disfavouring African workers).

The challenge that faces policymakers and planners today is therefore not necessarily to maximise mobility, or even to maximise its efficiency in cost or time (although those are desirable on their own terms). It is to consider how various parts of the South African economy – various fractions of both capital and labour – are favoured or disfavoured by the present system of mobility, and how any given intervention will do the same. This is not to say that all interventions are strictly zero sum – although some are – but rather that the benefits of any intervention are unevenly distributed, and have knock-on effects through the economy. More than that: the way people move is a crucial component of social regulation, and our reforms to transport need to be considered in that light.

In other words, we need to take the relatively simple (but not easy) challenge of facilitating faster and cheaper mobility across South African cities, and complicate it by understanding the place of mobility in South Africa's present and future mode of regulation. If our goal is a fundamentally more just growth regime, one where the balance of power and surplus is tipped more towards those who are currently excluded and exploited, then our transport and spatial policies need to be designed accordingly. Simply better transport systems alone will at best ameliorate and temper the injustice of South Africa's political economy, and its spatial form.

Acknowledgement

Many thanks go to Christina Culwick for her comments on an earlier draft. All errors that remain are mine alone.

References

Beinart, W. (2001) *Twentieth-Century South Africa*. Oxford: Oxford University Press.

Bester, R. (2001) City and citizenship. In: Enwezor, O. (ed.), *The Short Century: Independence and Liberation ovements in Africa 1945–1994*, 219–224. New York, NY: Prestel.

Bond, P. (2000) *Cities of Gold, Townships of Coal*. Trenton and Asmara: Africa World Press.

Bond, P. (2007) Introduction: Two economies – or one system of super exploitation. *Africanus: Journal of Development Studies* 37 (2), 1–21.

Boyer, R. (2010) Are there laws of motion of capitalism? *Socio-Economic Review* 9 (1), 59–81.

Boyer, R. and Saillard, Y. (2002) A summary of régulation theory. In: Boyer, R. and Saillard, Y. (eds.), *Regulation Theory: The State of Art*, 36–44. Hoboken: Routledge.

Crankshaw, O. (2008). Race, space and the post-fordist spatial order of Johannesburg. *Urban Studies* 45 (8), 1692–1711.

Cross, C. (2001) Why does South Africa need a spatial policy? Population migration, infrastructure and development. *Journal of Contemporary African Studies* 19 (1), 111–127.

Cross, C. (2013) Delivering human settlements as an anti-poverty strategy: Spatial paradigms. In: Pillay, U., Hagg, G. and Nyamnjoh, F. (eds.), *State of the Nation: South Africa 2012–2013*. Cape Town: HSRC Press.

Dawson, A. (2006) Geography of fear: Crime and the transformation of public space in post-apartheid South Africa. In: Low, S and Smith, N. (eds.), *The Politics of Public Space*, 123–142. New York and London: Routledge.

Diphoorn, T. (2015) The private security industry in urban management. In: Haferburg, C. and Huchzermeyer, M. (eds.), *Urban Governance in Post-Apartheid Cities: Modes of Engagement in South Africa's Metropoles*, 197–209. Pietermaritzburg: University of KwaZulu-Natal Press.

Freund, B. (1991) South African gold mining in transformation. In: Gelb, S. (ed.), *South Africa's Economic Crisis*, 4–7. Oxford: Oxford University Press.

Gelb, S. (1987) *Economic Crisis in South Africa: 1974–1986*. Unpublished seminar paper No 214. Johannesburg: University of the Witwatersrand African Studies Institute.

Gelb, S. and Saul, J. (1981) *The Crisis in South Africa*. New York, NY: Monthly Review Press.

Goodlad, R. (1996) The housing challenge in South Africa. *Urban Studies* 33, 1629–1646.

Greenberg, S.B. (1987) Ideological struggles within the South African state. In: Marks, S. and Trapido, S. (eds.), *The Politics of Race, Class and Nationalism in Twentieth Century South Africa*, London : Longman.

Hindson, D. (1991) The restructuring of labour markets in South Africa: 1970s and 1980s. In: Gelb, S. (ed.), *South Africa's Economic Crisis*. Oxford: Oxford University Press.

Isaacs, G. (2015) *The commodification, commercialisation and financialisation of low-cost housing in South Africa*. Working Paper No. 200, Leeds: FESSUD.

Jessop, B. (1988). Regulation theory, post Fordism and the state: More than a reply to Werner Bonefield. *Capital and Class* 12 (1), 147–168.

Khan, F. and Ambert, C. (2003) Preface. In: Khan, F. and Thring, P. (eds.), *Housing Policy and Practice in Post-Apartheid South Africa*. Sandown: Heinemann.

Leibbrandt, M., Levinsohn, J. and McCrary, J. (2005) *Incomes in South Africa Since the Fall of Apartheid*. National Bureau of Economic Research Working Paper. National Bureau of Economic Research: Cambridge, MA.

Lemanski, C. (2004). A new apartheid? The spatial implications of fear of crime in Cape Town, South Africa. *Environment and Urbanization* 16 (2), 101–112.

Lemanski, C. (2011) Moving up the ladder or stuck on the bottom rung? Homeownership as a solution to poverty in urban South Africa. *International Journal of Urban and Regional Research* 35 (1), 57–77.

Lewis, D. (1991) Unemployment and the current crisis. In: Gelb, S. (ed.), *South Africa's Economic Crisis*. Cape Town: David Phillip Publishers.

Lipton, M. (1988) Capitalism and Apartheid. In: Lonsdale, J. (ed.), *South Africa in Question*. Portsmouth, NH: Heinemann.

Lundahl, M. (1989). Apartheid: Cui bono? *World Development* 17 (6), 825–837.

Mabin, A. (1992) Comprehensive segregation: The origins of the group areas act and its planning apparatuses. *Journal of Southern African Studies* 18 (2), 37–41.

Mamdani, M. (1996) *Citizen and Subject: Contemporary Africa and the Legacy of Late Colonialism*. Princeton: Princeton University Press.

Marais, H. (2010) *South Africa Pushed to the Limit: The Political Economy of Change*. Cape Town: UCT Press.

McCarthy, J. (1992) Local and regional government: from rigidity to crisis to flux. In: Smith, D. (ed.), *The Apartheid City and Beyond: Urbanization and Social Change in South Africa*. London and New York, NY; Johannesburg: Routledge; Witwatersrand University Press.

Morris, M. (1991) State, capital and growth: The political economy of the national question. In: Gelb, S. (ed.), *South Africa's Economic Crisis*. Cape Town: David Phillips Publishers.

Neilson, D. (2011). Remaking the connections: Marxism and the French regulation school. *Review of Radical Political Economics* 44 (2), 160–177.

O'Meara, D. (1996) *Forty Lost Years: The Apartheid State and the Politics of the National Party: 1948–1994*. Ohio: Ohio University Press.

Perelman, M. (2000) *The Invention of Capitalism*. Durham and London: Duke University Press.

Perelman, M. (2007) Articulation from feudalism to neoliberalism. *Africanus: Journal of Development Studies* 37 (2), 22–38.

Pirie, G. (1992) Traveling under apartheid. In: Smith, D. (ed.), *The Apartheid City and Beyond: Urbanization and Social Change in South Africa*. London and New York, NY; Johannesburg: Routledge; Witwatersrand University Press.

Salazar Ferro, P., Behrens, R., & Wilkinson, P. (2013). Hybrid urban transport systems in developing countries: Portents and prospects. *Research in Transportation Economics*, 39 (1), 121–132.

SA Reserve Bank (1991) *Quarterly Bulletin: September*. Pretoria: South African Reserve Bank.

SAIRR (1992) *Race Relations Survey 1991/92*. Johannesburg: South African Institute for Race Relations.

Sihlongonyane, M.F. (2015) A critical overview of the instruments for urban transformation in South Africa. In: Haferburg, C. and Huchzermeyer, M. (eds.), *Urban Governance in Post-Apartheid Cities: Modes of Engagement in South Africa's Metropoles*. Stuttgart: Gebr. Borntraeger Verlagsbuchhandlung.

Smith, D. (1992) Introduction. In: Smith, D. (ed.), *The Apartheid City and Beyond*. London: Routledge.

Todes, A. (2006) Urban spatial policy. In: Pillay, U., Tomlinson, R. and Du Toit, J. (eds.), *Democracy and Delivery: Urban Policy in South Africa*. Pretoria: HSRC Press.

Todes, A. (2015) The external and internal context for post-apartheid urban governance. In: Haferburg, C. and Huchzermeyer, M. (eds.), *Urban Governance in Post-apartheid Cities: Modes of Engagement in South Africa's Metropoles*. Stuttgart: Gebr. Borntraeger Verlagsbuchhandlung.

Turok, I. (2001) Persistent polarisation post-apartheid? Progress towards urban integration in Cape Town. *Urban Studies* 38 (13), 2349–2377.

Vidal, J-F. (2001) Birth and growth of the regulation school in the French intellectual context (1970–1986). In: Labrousse, A. and Weisz, J-D (eds.), *Institutional Economics in France and Germany*, Heidelberg: Springer.

Walker, O. (1948) *Kaffirs are Lively*. London: Gollancz.

Wolpe, H. (1995) Capitalism and cheap labour power in South Africa: From segregation to apartheid. In: Beinart, W. and DuBow, S. (eds.), *Segregation and Apartheid in Twentieth-Century South Africa*. London: Routledge.

Wood, A. (2015) Transforming the post-apartheid city through bus rapid transit. In: Haferburg, C. and Huchzermeyer, M. (eds.), *Urban Governance in Post-Apartheid Cities: Modes of Engagement in South Africa's Metropoles*, 79–98. Pietermaritzburg: University of KwaZulu-Natal Press.

Worden, N. (2000) *The Making of Modern South Africa: Conquest, Segregation, and Apartheid*. Oxford: Blackwell.

9 Constructing wellbeing, deconstructing urban (im)mobilities in Abuja, Nigeria

Daniel Oviedo, Caren Levy and Julio D Dávila

This chapter engages with a discussion on the role of transport mobility in the wellbeing of low-income urban citizens of the Global South. Using the case of Abuja, Nigeria's capital city, we explore transport-related vulnerabilities and disadvantages for low-income and vulnerable groups and how they are addressed by local transport. First, the chapter aims to illustrate the importance of new knowledge(s) and methodologies critical to the future of transport planning in the Global South. Moving away from traditional measures of mobility, the chapter explores the potential of wellbeing as an operational concept in transport planning. By focusing on the different dimensions of wellbeing, a more rounded view of transport in the life of urban dwellers is developed, which leads transport planning down new avenues of knowledge and methodologies in the pursuit of more socially just cities. Second, in doing so, the chapter also seeks to reflect on what are essential mobilities and to explore the contribution of transport to achieve a quality of life that recognises the diverse identities of all urban citizens. The chapter showcases evidence of transport-related wellbeing in three dimensions: material, relational and subjective, building on information of daily travel practices, social and cultural identities, socio-demographic characteristics and subjective perceptions from low-income and vulnerable populations. Such information was collected through quantitative and qualitative instruments tailored to a conceptual framework for understanding personal wellbeing. Findings confront objective and subjective measures of wellbeing, suggesting added relevance of transport as either a potential enabler or constraint to personal autonomy and freedom, as well as the relevance of security and personal and collective expectations in defining the influence of transport policy in the lives of lower-income citizens.

Introduction

The current crisis in mobility in most large cities of the Global South calls into question the entire traditional transport planning endeavour: its ideological underpinnings, its institutional framing and its methodological practices. The primary focus on private motorized vehicles, the speed/time measures of success and the development of transport provision in the interests of the few are obsolete in the face of widespread urban congestion and socio-spatial inequality. The links between congestion and inequality are complex, made up of the interaction between structural factors related to the political economy and socio-cultural processes, and urban practices related to governance, a range of planned interventions and the everyday struggle of urban citizens to live a decent life.

Addressing this crisis in physical and socio-economic terms requires rethinking some of the tenets of transport planning. This chapter is a contribution to this discussion in two ways. First, it aims to illustrate the importance of new knowledge(s) and methodologies critical to the future of transport planning in the Global South. Moving away from traditional measures of mobility, the chapter explores the potential of wellbeing as an operational concept in transport planning. By focusing on the different dimensions of wellbeing, a more rounded view of transport in the life of urban dwellers is developed, which leads transport planning down new avenues of knowledge and methodologies in the pursuit of more socially just cities. Second, in doing so, the chapter also seeks to reflect on what are essential mobilities and to explore the contribution of transport to achieve a quality of life that recognizes the diverse identities of all urban citizens.

This is explored in the case of Abuja, Nigeria's capital, and arises from a research project on transport in informal neighbourhoods in Abuja, Kaduna and Ibadan, three of Nigeria's largest cities. The research draws on primary qualitative information from semi-structured interviews and quantitative information from transport-user surveys collected in 2015 by a team from the Development Planning Unit (DPU), University College London (UCL), working with Nigerian partners.[1]

The focus on Abuja must be viewed in the context of urbanization and the development of planning before, during and after colonization in Nigeria. Linked to the development of powerful empires in the North and the South of the country in the eighteenth and nineteenth centuries, Nigeria has a long history of urbanization that pre-dates the colonial period (Mabogunje, 1968 in Solanke, 2013). Colonization brought with it new patterns of mobility because of accelerated urban growth and a change in spatial patterns associated with the development of roads and railways to serve the extractive practices of the colonial economy (exports of local natural resources such as minerals). Colonial approaches to urban development in the early twentieth century were largely predicated on 'the sanitation syndrome' (Swanson, 1977 in Mabogunje, 1992), a common impetus driving colonial administrations worldwide (on Bombay, India, see McFarlane, 2008). These early planning approaches were 'racist in orientation and segregationist in practice' (Mabogunje, 1992: 74) and in Nigeria developed into a three-sector division of cities, reflecting different degrees of intervention based on whether there were European settlement in cities or not (Home, 1983; Falola, 1996). The impact of that period was not only on the differential and unequal engagement with governance, the management of land and provision of infrastructure and housing in the development of particular cities, but also on urban development concentrated in a handful of cities that served the interests of the colonial economy (Falola, 1996).

With independence from Britain in 1960, the aspirations of the new Nigerian government were to address both the planning and management of cities, as well as to plan a national urban system that 'no longer serves the purpose of an outward-oriented colonial economy, which is less primate in its size distribution, and whose locational distribution within the national territory conduces to even development in the country' (Mabogunje, 1990: 146). Combined with a political determination to create a balanced federal system in such an ethnically diverse country, Nigeria went from three regional administrations in 1960 to 21 states by 1986. The capital cities of these states have served as nodes of high urban growth (Mabogunje, 1992; Home, 1983; Falola, 1996), where the interaction between urban development and the provision of road transport, particularly in peri-urban areas, has been important in the commodification of land and city expansion. Abuja enters the picture in 1976, when it was declared the new capital of Nigeria, a distinct federal capital territory of its own, planned around a modernist urban vision in a post-independence era.

In the next sections the definition of wellbeing as it relates to transport is discussed. This is followed by a more in-depth examination of transport and urban development in Abuja, and then a section drawing on research findings related to different dimensions of wellbeing and mobility in Abuja. The conclusions review the lessons learnt.

Why wellbeing and transport mobility?

Rationale and definitions

In urban settings in the Global South, lack of transport can become an obstacle to full participation in society, economic development and overall improvement in quality of life. Wachs (2010) argues that transport is central to poverty alleviation because it enables access to essential opportunities and social interactions. Poor households suffer measurable deficits in nutrition, health, education and opportunities to work, which are invariably correlated with deficits in mobility (Titheridge et al., 2014) and which are also often correlated with political marginalization. An intersectional view of poor households further highlights differential deficits on the basis of gender, age, disability and other social relations (Levy, 2013; Jones and Lucas, 2012; Oviedo and Titheridge, 2015). The built environments of cities in the Global South are characterized by their comparatively undeveloped road infrastructure, urban primacy and spatial inequalities (Cervero, 2013), which can restrict the ability of urban populations to appropriate the city and make full use of its opportunities. International agendas largely concerned with infrastructure development and mobility improvements have exacerbated environmental degradation and unequal distribution of opportunities arising from differences in the distribution of urban investments and people's capacity to travel (UN-HABITAT, 2013). Despite these challenges, large-scale transport interventions, often driven by technocratic approaches to urban planning and development, focus mostly on increasing speed of travel and maximizing passenger movement but rarely incorporate explicit aims such as social equity or increased wellbeing.

The notion of wellbeing is increasingly used to assess the impact of development policy (Boarini et al., 2006; Australian Government, 2010; and ONS, 2014). In many countries, there is a growing aspiration for incorporating wellbeing into policy goals and appraisal (Manderson, 2005). However, the multidimensional nature of the concept constrains its operationalization and feeds concerns on its validity for sector-specific assessments. In the face of these challenges, researchers tend to focus on specific dimensions and descriptions of wellbeing, making it difficult to reach conceptual and methodological agreements (Dodge et al., 2012; La Placa et al., 2013).

Transport has the potential to increase wellbeing through increased access to opportunities essential to lead a decent life, reduction in vulnerability to externalities and the promotion of active travel (i.e., walking and cycling) under appropriate conditions of safety and comfort. In cities of emerging economies, some of these objectives may be secondary in transport policy and development. There is the potential to integrate questions of wellbeing and inequality into transport policy and investment so as to more effectively redress transport gaps. And yet, despite contributions from different perspectives suggesting links between mobility and wellbeing, detailed examination of their relationships is relatively recent.

Definitions of wellbeing in transport research often focus on access to assets and connections required to meet needs, as well as on physical and mental health. Most conceptual and methodological advances have leaned towards health-related and subjective dimensions (Vella-Brodrick, 2011). This may have been influenced by a large body of research over the past 20 years that has dealt with issues such as accessibility, transport deprivation and social exclusion as part of the examination of the social dimensions of transport, (see Van Wee and Geurs, 2011; Jones and Lucas, 2012; Levy, 2013).

Research in industrialized societies analyzed the effects of mobility on the wellbeing of the elderly (Banister and Bowling, 2004; Mollenkopf et al., 2005; Spinney et al., 2009), with a more recent shift towards the effects of transport on notions such as connectedness, relatedness and satisfaction with life (Pacione, 2003; Delbosc and Currie, 2011; Stanley et al., 2011; Vella-Brodrick, 2011). Evidence from recent research on transport disadvantage and transport-related social exclusion (Stanley and Lucas, 2008; Stanley and Vella-Brodrick, 2009; Delbosc and Currie, 2011; Vella-Brodrick and Delbosc, 2011) suggests that the mainstream goals of transport planning of connectedness and accessibility arise from an inadequate understanding of the relevance of transport in human life, pointing to wellbeing as the potential missing link in the development of transport policies and practices.

One of Nigeria's greatest challenges is to reconcile the policy objective of increasing the wellbeing of its citizens while seeking to promote economic growth, an aim that in many contexts has led to greater concentration of wealth and not necessarily to a concurrent reduction in poverty or improved wellbeing among the majority of the population. By linking multiple dimensions of wellbeing with urban transport, we intend to provide new empirical and conceptual elements to help inform this reconciliation into urban transport policy.

Under the premise that transport is a multidimensional construct, and building on empirical and conceptual evidence showing that mobility is an integral part of socio-economic and political disadvantage (Ohnmacht et al., 2009; Lucas, 2012), our research offers a framework to link transport and wellbeing. This is a person-centred framework, with three interdependent dimensions, material, relational and subjective, as shown in Figure 9.1. The material aspect of wellbeing encompasses the economic realm, but also incorporates elements of 'human capital' (Becker, 1962) such aseducation and health, as well as characteristics of the physical and natural environment. This dimension includes both the material assets required for attaining an adequate standard of living and the set of skills, abilities and environmental conditions to secure such assets and so be able to 'live well'. This approach to assessing the required assets and opportunities for a full life is strongly aligned with the concept of human development, which has governed recent development policy in the Global South. Human development is defined as the process of enlarging people's freedoms and opportunities and improving their wellbeing (UNDP, 1990). Such freedom encompasses the opportunities people have to decide who to be, what to do and how to live. The ideas around human development have shaped significant breakthroughs in international development policy that span from the publication of the report by Stiglitz et al., (2009) where they redefine economic and social progress, to the definition of the current Agenda for Sustainable Development for 2030 agreed in September 2015 at the General Assembly of the United Nations (van Lindert, 2016). This approach also provides ample space for freedom, which responds to the pluralism that defines the possible ways of living developed by every person in every context (Alkire, 2015).

The second dimension refers to the relational aspect of adequate living. Personal and social relations are central, incorporating social capital and relational experiences that contribute to a full life. This refers to relations of love and care and networks of support and obligation, the potential to develop social, political and cultural identities, and to take collective action within supportive governance structures, in the context of safety and security. The freedom to express these relates directly to the inherent feature of transport as public space, while other aspects of these elements involve complex social and personal characteristics where mobility may be less directly implicated than in the material dimension.

The third dimension relates to subjective perceptions, values and experiences (White, 2010). This is one of the most studied dimensions of wellbeing, arising from psychological literature (Vella-Brodrick, 2011). This encompasses perceptions of people's surroundings and their own life in a larger moral and social context, and also in a public and private spatial context that fosters the experience and expression of valued subjective characteristics. Wellbeing in this sense is often equated with 'happiness', leading to several indices aiming to measure life satisfaction and happiness as proxies to a full life (Camfield and Skevington, 2008).

Methodological approach

The methodological challenges of adapting this framework to transport involve producing tangible and interrelated evidence on the dynamic relations between the different dimensions of wellbeing and transport. We adopt a mixed methods approach to gathering evidence (see Figure 9.1).

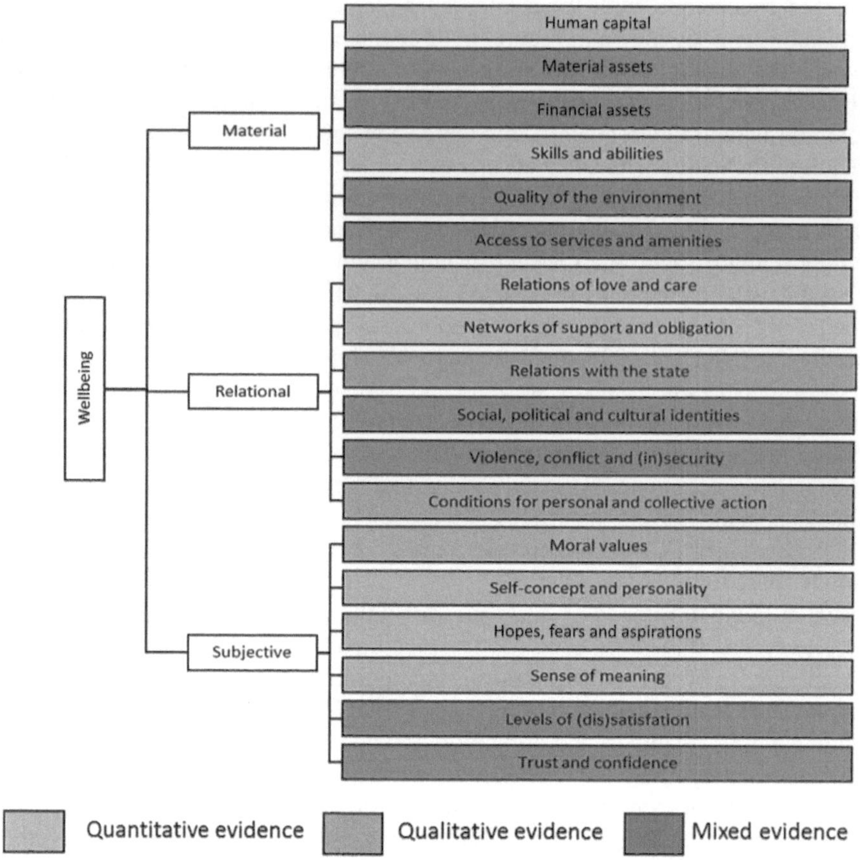

Figure 9.1 Transport and wellbeing: conceptual framework and type of evidence

Source: Own elaboration based on White (2010).

From a quantitative angle, data were collected through a field survey of 337 individuals in three low-income settlements in Abuja's metropolitan region (Jahi Village, Galadimawa and Kabusa – Figure 9.2). This involved an 'interception survey' whereby individuals waiting for transport services were approached with a set of questions. Doing this sought to: 1) identify commuters' residential location and demographical and socio-economic characteristics; 2) capture travel behaviour, travel choices and stated travel needs; and 3) identify perceptions of availability of transport choice, transport difficulties, access to activities and the impact of transport on quality of life and on accessing opportunities for economic and personal development.

The surveyed population is composed largely of young adults aged 20 to 40 (93%), with males dominating the sample (59%). This is an important feature to bear in mind, because it could significantly shape the interpretation of perceptions,

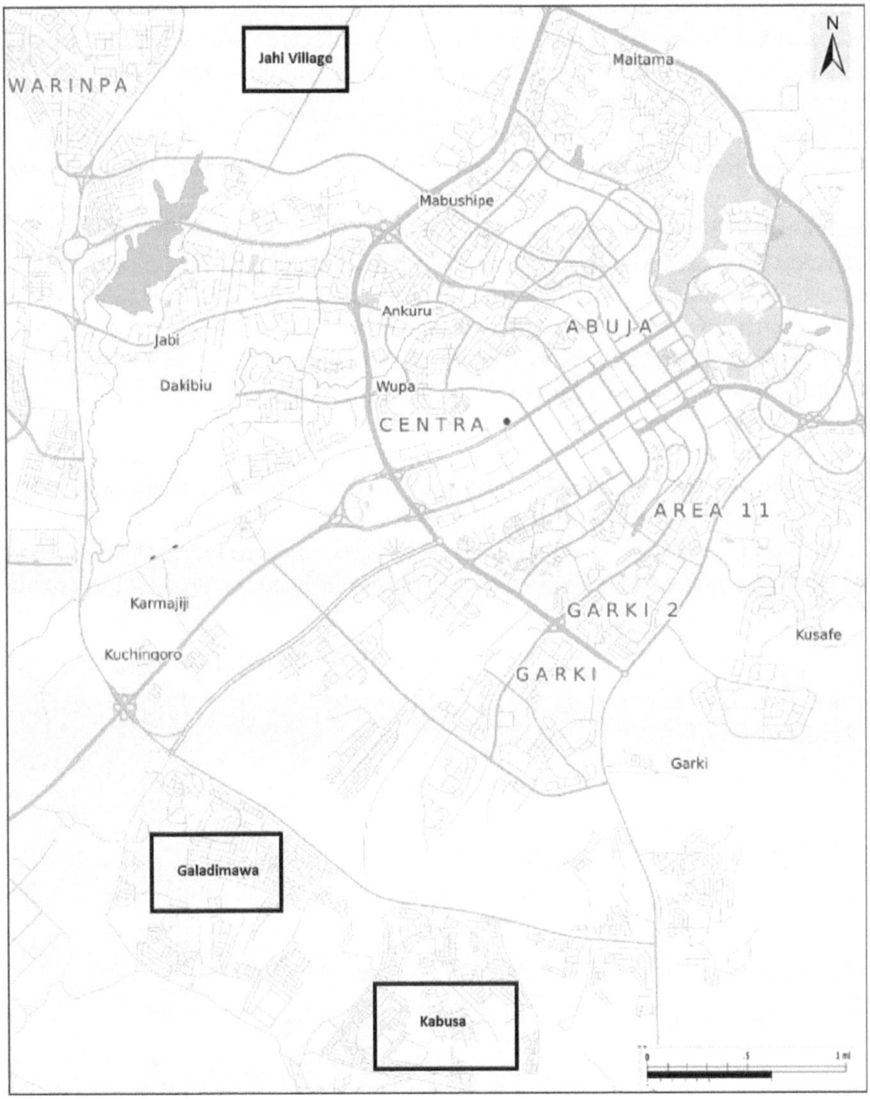

Figure 9.2 Abuja: location of interception transport survey

Source: Own elaboration.

practices and vulnerabilities. For example, lower presence of women in public spaces could reflect lower rates of labour-market participation or arise from social conventions restricting their circulation. Also of note is the fact that three quarters of respondents are Christian, with the remaining quarter stating that they are Muslim.

Income is a major influence on wellbeing. Socially vulnerable groups such as low-income populations are more susceptible to the negative externalities of urban transport and thus could be more deeply affected. Nearly half of respondents (48%) reported earnings below the national minimum wage of 18,000 Naira/month (US$90) (The Nation, 2015), with a larger comparative share of women in the lower-income bracket than men (46% of women compared with 42% of men).

The relational dimension was examined through semi-structured interviews with 18 residents in total between the selected settlements in Abuja, and focused on how people interact with transport and its public-space character, provision of transport services and interaction with the state through transport. The subjective dimension permeates most of the information collected in the survey, since most of the variables have a subjective element, some more than others. The issue of freedom to move around and therefore the reliability of the system is very important for respondents. Because of limitations inherent to the survey, additional information was compiled to help shed light on planning and practice, challenges faced by current policies to increase wellbeing of urban populations through transport, and governance issues that might limit people's ability to influence transport services so they contribute more effectively to their wellbeing.

In this vein, a series of in-depth interviews was conducted (Figure 9.3) with three types of respondents identified with the help of our local research partner: public

Figure 9.3 Data collection in Abuja

Photo: Odukogbe, 2015.

officials involved in urban transport planning and development, local academics and transport service providers. Seven interviews were conducted in Abuja.

Abuja: a new capital still under construction

Abuja is a post-colonial city. Created in 1976, the new federal capital sought 'to stand not only as a symbol of a new national consciousness but also as a vignette of the vision of the African city entertained if not by African planning then certainly by African decision makers' (Mabogunje, 1990: 146). International consultants submitted the Master Plan (1976–2000) to the Nigerian government in 1979. A Federal Capital Development Authority (FCDA) was set up to implement the Master Plan, and construction started in 1981. In 1985, the FCDA was brought under the authority of the Ministry of the Federal Capital. In 1991, the seat of government was moved there from Lagos, the main port and national commercial centre (Femi, 2012; Jibril, 2015).

The symbolism of Abuja goes beyond its power as a statement of pride and unity in a post-independence state. It was to be a counterpoint to Lagos, seen as a city in crisis, as reflected in traffic congestion and a failed system of land and urban development (Adeponle, 2013; Femi, 2012). Conceived as a crescent or horseshoe, Abuja's Master Plan was 'intended to regulate land use, transportation systems, infrastructure, housing and other services in a manner that recognized their interrelationships and spatial requirements' (Adeponle, 2013: 147). The Master Plan was to be developed in four phases over 25 years.

Contemporary Abuja differs from the Master Plan proposals in several ways that are interrelated and relevant to the current transport situation. First, the Plan proposed a maximum population of 3 million by 2000, with growth over and above this to be housed in designated satellite towns (Jibril, 2015). It is estimated to have currently a base population of more than 3 million people, but with the daily influx of commuters, its daytime population can rise to 7 or 8 million (Iro, 2007; Femi, 2012). A magnet for in-migration, at more than 8% per annum, Abuja has one of the fastest population growth rates in Nigeria and among the fastest in Africa (Myers, 2011). Growth is faster in satellite settlements, which according to Abubakar and Doan (2010), are growing at about 20% per annum.

Second, while the Master Plan proposed a new city, the chosen location was not empty of people. It required the relocation of 845 villages (Adeponle, 2013). However, over time and driven by different political regimes, there has been some policy vacillation about their resettlement (and compensation) or integration into the Plan (Jibril, 2015). The outcome has been, on the one hand, the forced evictions of more than 800,000 people between 2003 and 2007 (Fowler et al., 2008) and on the other, along with the lack of planned housing provision, the growth of informal settlements in the city (ibid.; Adeponle, 2013).

Third, implementation has not advanced as planned. Phase 1 is largely completed, focusing mainly on government and private development. Subsequent phases demonstrate progressively less completion from phases 2 through 4 in which the lack of planned housing, infrastructure and transport are the most critical aspects. Nevertheless, the population continues to grow, so 'the present reality today shows an increase in the size of the FCC from the original 250 square kilometers to 1,123 square kilometres' (Jibril, 2015: 14). The result is acute housing shortages (Abdullahi and Aziz, 2010; Umoh, 2012), proliferation of informal settlements (Amba, 2010; Jibril, 2005), occupation of uncompleted buildings (Abubakar and Doan, 2010), inadequate supply of water and sanitation to the poor (Ilesanmi, 2006: Ojo, 2011) and traffic congestion (Abubakar, 2014).

Transport infrastructure was central to structuring the Master Plan, with the original vision for the city one of' . . . wide green belts separating neighborhoods of different social classes . . . low-density and low-rise buildings with wide streets based on the expectation of the highly motorized population. . .' (Mabogunje, 1992: 147). Each phase was bounded by large-scale expressways, complemented by the presence of a road-connected system of satellite towns such as Idu (the major industrial site) and Dei-Dei (home of the international livestock market and the international building materials market). However, whilst road provision is 90% complete in Phase 1, it is only respectively 35%, 25% and 5% completed in subsequent phases (Femi, 2012).

The plan's pledge for a "provision of public transport mobility to residents who do not own cars" (Femi, 2012: 120) has not been implemented. This has resulted in congestion and the expansion of alternative informal-sector mass transit modes, and in increasing tensions among transport providers, as well as with Federal Capital Territory (FCT) bodies. Rail infrastructure planned in Phase 1 (Jibril, 2015) is only now being provided in the form of a light rail system (LRT). Similarly, a bus rapid-transit system (BRT) is being implemented, though without inclusive citizen participation and coupled with a ban in 2012 of commercial motorcyclists and mini-bus operators (Ebo, 2015). Some suggest that the current transport difficulties arise from the separation of transport policy under the auspices of the Transport Secretariat of the FCDA, and the provision of infrastructure in another part of the FCDA (Femi, 2012).

As shown in the next section, the result is that most motorized transport users resort to a range of informal and non-motorized transport modes.

Mobility in Abuja

This section presents the evidence on the transport-wellbeing relationship in its three dimensions: material, relational as an aspect of life and perceived, as applied to the three low-income settlements in Abuja (See example in Figure 9.4) that are the focus of the case study.

The material dimension of wellbeing: individual and collective assets and accessibility

A vast informal sector exists in Abuja, which currently employs two thirds of residents in the municipality and close to 80% in the larger FCT (FCT MDG Office, 2009). Acknowledging that this was a long-term reality in Nigeria, the Master Plan projected that 40% of jobs would be of an informal nature, and allowed space

Figure 9.4 Jahi Village, Abuja
Photo: Odukogbe, 2015.

for informal businesses in designated locations (IPA, 1979; Abubakar, 2014). However, small-scale traders are often unable to afford rents, pushing them to hawking or using roadsides to sell their goods and services, with implications for daily mobility. Similar levels of informality are also present in transport services, particularly in satellite settlements such as those in this research. It is striking that most survey respondents use informal transport modes such as *okada* (motorcycle taxis), *danfo* (minibuses), *keke napep* (rickshaws) and shared taxis (56% of the total sample). A total of 57% of female respondents resort to these modes, with *okada* the most frequent one, while 70% of respondents who said to be unemployed also use informal modes. Non-motorized transport represents the second largest category, with higher dependency on this among certain social groups. For example, 37% of women complete most of their trips on foot, and between 34% and 40% of people in the three lowest income ranges also walk for most of their trips.

A quadrant analysis (Currie and Delbosc, 2011) shows that for a significant share of the sample (39%) it is both difficult and important to access transport without depending on others (Figure 9.5). Furthermore, for 55% of the sample this is an important feature of their mobility, which can be interpreted as a potential difficulty in their daily travel practices.

A substantial majority (79%) of those surveyed said that they had difficulties in accessing productive activities. Lower-income respondents are mainly users of non-motorized and informal means of travel, which tend to be slower, more polluting and more uncomfortable than formal modes, with knock-on effects on users' physical wellbeing and availability of time for other activities. However,

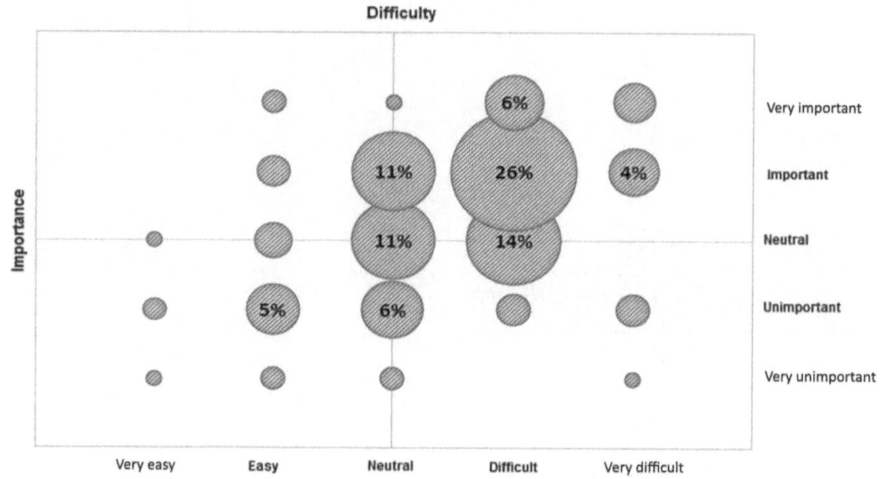

Figure 9.5 Using transport without depending on others (% of respondents)

Source: Own elaboration based on field data.

people reporting difficulties also have a higher mean income, which also has implications for the location of such opportunities. This calls for further reflection regarding expectations of different social groups and the subjective nature of wellbeing analysis. It also requires higher considerations for decision-making in balancing mobility conditions for different social groups and increasing expectations of low-income populations.

A closer examination of the data summarized in Table 9.1 reveals that people who do not perceive difficulties in accessing transport for productive activities benefit from a larger local supply of informal transport and tend to walk less, despite having lower average incomes, while tending to travel more to other satellite areas rather than to the centre of Abuja. An analysis of the average percentage of individual monthly income spent on a single trip shows two dominant categories of expenditure: those who spend 1% of their income or less (39% of the sample), and those who spend more than 5% (34% of the sample). Among respondents in the first category, the share of non-motorized travel is approximately 50% of all trips. A higher proportion of male respondents said they experienced difficulties, a perception that can be partly explained by the fact that men tend to perceive themselves as breadwinners in the household.

Traditional transport analyses rarely explore accessibility beyond travel to work. Although other trips such as leisure, retail and visits to friends and family are very relevant for material wellbeing, interpreting this dimension involves availability of opportunities to allow full participation in society and a rewarding life depending on individual interests and priorities. For 64% of respondents, employment is not easily available, rating it as average or below. By contrast, a

Table 9.1 Perceived difficulties in accessing economic opportunities because of transport

Factor	*Economic opportunities*		*Health and education*		*Religious facilities*	
	Yes	*No*	*Yes*	*No*	*Yes*	*No*
Share of respondents	79%	21%	47%	53%	20.50%	79.50%
Mean income (US$)	144.4	104.1	143	128	137.4	136
Average age	30.05	31.39	29.2	31.3	30.5	30.3
Mean level of education	Secondary	Secondary	Secondary	Secondary	Secondary	Secondary
Gender	69% Male	58% Male	55% Male	76% Male	68% Male	67% Male
Mean trip cost/ income	1.40%	2.60%	1.80%	1.50%	2.10%	1.50%
Availability of opportunities	3.1	2.64	3.1	2.8	2.64	3.09
Travel cost score (out of 5) *	3.21	2.92	3.39	3.31	3.35	3.35
Travel time score (out of 5) *	3.29	3.01	3.48	3.33	3.26	3.44

*Availability of employment is measured on a scale from 1 to 5 where 1 is very low, 3 is indifferent and 5 is very high

Source: Own elaboration based on field data

large majority believe that education, shops and leisure are easily available at the city-scale, suggesting that transport may be facilitating their access to these.

The ability to get on and off a means of transport is an important feature for most respondents and is related both with the physical conditions of people and transport features. Nearly 60% find difficulty in getting on and off particular transport modes. This is also related with social identities and conventions. Women find it difficult to get on and off *okadas* and *keke napep*, which according to interviewees is related to the customary sitting position in the motorcycle taxis and the limitations of traditional clothing for riding this type of vehicle. Older adults also perceive these difficulties in their daily mobility, and it is one of the higher-scoring transport features among this group.

Another material dimension of wellbeing relates to transport affordability. Interestingly, despite the sample of respondents being overwhelmingly low-income when compared with average incomes and the legal minimum wage, affordability would not appear to be an explanatory variable in preventing users from gaining access to a range of opportunities. By contrast, the amount of time spent travelling scores negatively, showing that respondents place a premium on spending less time in transport. This might reflect the trade-off against cost for some of those walking and/or taking non-motorized modes of transport, which almost invariably take longer. Not surprisingly, higher purchasing power leads to better transport affordability, particularly in motorized modes. This entails class inequalities in the access to different types of transport and the ability of different social groups to appropriate and exercise a right to the city. Pockets of congestion and concentration of infrastructure and traffic can have localized effects that increase inequalities in the quality of the environment.

One of the main issues to consider in the material dimension of wellbeing is the quality of the natural and built environment. Residents in Abuja appreciate the fact that the main streets are laid out according to a plan, and regularly maintained. However, this view contrasts with the recommendations made by respondents when asked what could be done to improve the transport system, as reported in Table 9.2 (see the section on the subjective dimension later in this chapter). There, respondents highlighted quality of the road network as a number one priority for improvement. This can be interpreted as referring largely to neighbourhood streets, rather than the citywide network, given deficiencies in maintaining the local road network.

The relational dimension: access to social interactions, security and opportunities for full participation in society

The relational dimension of wellbeing was examined mainly through semi-structured interviews and focused on how people interact with transport, how transport is provided and how people interact with the state. Respondents perceive the state as enabling people to access transport to interact among themselves, practice their religious beliefs and use public space. A large share expressed dissatisfaction with the difficulties they face (see Table 9.1). For example, 20% of respondents said that they have transport difficulties in accessing religious facilities. Males are more acutely affected both in accessing religious facilities and reaching family members. When this is correlated with availability of religious activities, the score is negative (just over 2).

Table 9.2 Perceived priorities for public investment/action at the city level

Priorities for public investment/action	
	Percentage of responses
Better bus stop/interchange	1%
Better roads	39%
Cheaper transport fare	5%
Less crowded vehicles	2%
Less noise/air pollution	4%
More availability across city/more frequent services	2%
Other	8%
Pedestrian pavement footpath	8%
Reduced road congestion	3%
Street lights	4%
Traffic order/better compliance	24%

Source: Own elaboration based on field data

Evidence from the relational dimension shows that transport can play a significant role in enabling both mobility and the appropriation of space by citizens to facilitate their participation in social groups and interactions. Interestingly, no one reported difficulties in accessing their social connections. Adequate and reliable transport, especially at times when people socialize, can help increase relational wellbeing. The evidence also calls for complementarity between transport policies and policies from other sectors such as land use and public space.

Another important element to highlight is security in relation to crime and safety from road accidents. The availability of night transport is perceived as being both important for respondents and a source of difficulty. Most respondents feel that it is important that transport provides them with some security, particularly in relation to terrorism and crime. People feel more secure in Abuja than in the satellite locations. The availability of transport at times when crime is higher, or when people perceive higher vulnerability to it, is a relevant difficulty for more than 75% of the sample (Figure 9.6), with nearly 72% of respondents highlighting availability at night as important. This can be a hindering element for individual freedoms to move around and meet people. Supply under adequate conditions of security is a responsibility of the state and can be complemented with availability of services and infrastructure that can increase perception of security at night.

Crime and terrorism can govern travel practices and access to opportunities, relations and overall satisfaction of transport. Young women are particularly concerned about protection from crime, and they find it more difficult to feel safe when using transport. There are significant differences as to how people with

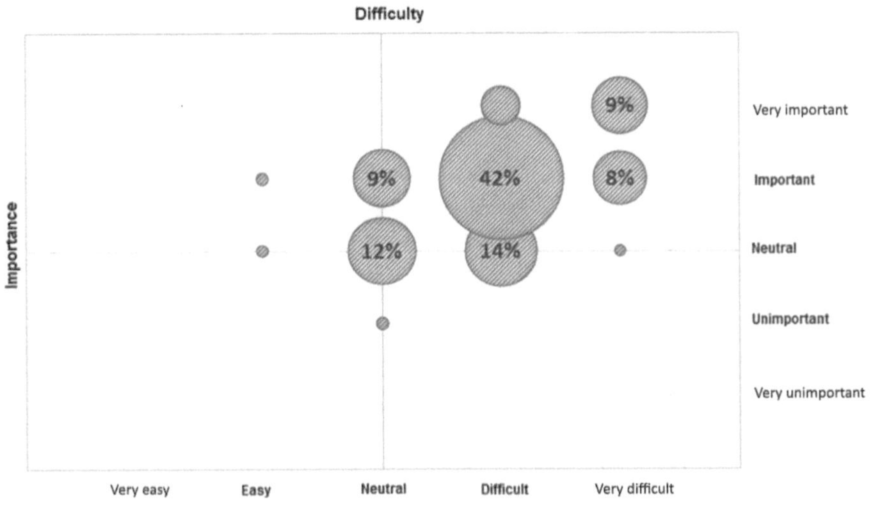

Figure 9.6 Transport availability at night (% of respondents)

Source: Own elaboration based on field data.

different social identities perceive terrorism. Of the 24% for whom feeling safe from terrorism is both important and difficult, nearly 65% are women. The mean score of the perception of safety from crime at the neighbourhood level (3.24 out of 5) is lower than at the city scale (3.63 out of 5). This reveals the relevance of crime as a hindering factor for local mobility and the relevance of urban infrastructure and facilities at the local level that can increase perceptions of security and reduce difficulties in feeling safe in daily travel practices. The higher city-scale score could be related to better quality of the built environment and greater presence of security services resulting from the high concentration of public agencies, including national ministries, and foreign diplomats.

In relation to traffic safety, there is a strong perception that traffic is chaotic so that the second most frequent recommendation for authorities is to enforce better compliance with rules (see Table 9.2 in the following section). This is at the same time a relational and subjective dimension, which is worth highlighting and exploring further.

What emerged from interviews with public officials in departments responsible for both the long-term planning and day-to-day operation of Abuja's public transport is that there are entrenched institutional challenges to improving planning and relationships with the population. At the core of this is the lack of a metropolitan authority for managing transport in the FCT. Transport planning is one of many central government functions, and is therefore more susceptible to slow bureaucratic procedures and political meddling. A bill recently brought to the National Assembly for the creation of a metropolitan authority for the FCT that resembles the structure of Lagos' Metropolitan Transport Authority (LAMATA) has been under consideration for more than two years, with poor prospects, partly because of limited political support. Without the legal authority and associated legitimacy, many technical decisions can be overturned by political interests.

According to interviewees, conditions are not conducive for building and consolidating capacity within public agencies involved in urban transport planning and management. There is a need to put a 'structure in place to attract talent currently being drawn to the private sector or going abroad' (M, 39, CS, MA).[2] Without strong institutions, it will become very difficult to assess impacts of transport developments, constraining the understanding of local officials and planners of the priorities for improving transport, particularly for Abuja's poor. A stronger conceptual approach is required to understand the contribution of transport to wellbeing that can counterbalance traditional economic rationale underpinning decision-making in transport planning. Such traditional rationale is closely connected to the evidence of priorities from the perspective of public officials: higher advocacy for infrastructure development for transport; development of a feeding routes programme for the BRT and LRT projects; redefinition of subsidies currently in place to reach target populations; and integration with different forms of formal and informal transport, including providers of interstate transport services. These largely technical issues contrast with the quantitative evidence collected in the case studies, revealing the need for more frequent and open interactions between transport planners and decision-makers on the one hand and the population on the other, particularly in satellite settlements. Development of an integrated approach to wellbeing that can inform transport policy can contribute to positive change in the approach from planners and decision-makers to transport investments and development, recognizing the limitations of transport interventions in addressing multiple sub-dimensions of wellbeing discussed in this chapter.

The subjective dimension: freedom, satisfaction and autonomy

The subjective dimension is the third element in our framework. For respondents, the freedom to move around, and therefore the reliability of the system, are very important. A general perception is that the transport system fails to provide them with a reliable means of moving quickly between places at times that suit them, thus restricting their autonomy. Among 19 respondents (6% of the total) there is a belief that the transport system has prevented them from accessing key places, such as the city centre, or has not allowed them to establish and maintain social interactions. Although small compared with the sample size, it is nonetheless a worrying finding. This relates to earlier concerns regarding the subjective perceptions of people in different social positions about what they can expect from transport investments. Results consistently suggest that lower-income and other socially vulnerable populations have lower expectations than better-off groups; targeted considerations in transport policy and decision-making are required to deliver a more equitable response to differentiated needs and to balance future expectations.

The issue of autonomy and liberty to move in urban space can be related to difficulties with spatial and temporal autonomy, affecting access to opportunities and enjoyment of life. These issues are perceived as being more important, though transport is perceived as having a limited ability to address subjective needs of reliability, independence and liberty. More than 64% of respondents find it difficult to reach places quickly with transport; 70% perceive difficulties in finding transport services and alternatives; 36% find difficulties in changing transport services; and 65% have difficulties moving at any time (Figure 9.7).

For respondents in Abuja, the city's transport system offers more spatial autonomy and reliability compared with the other two Nigerian cities examined in the larger research project, although freedom to move at different times of the day is perceived as a constraint. Weekends and festive seasons are frequent references to times when transport is insufficient to allow expected levels of travel. This is especially true in relation to social interactions; these periods are the most relevant for relational wellbeing because it is the time when people tend to socialize. Something similar occurs with the loss of spatial autonomy. This affects all dimensions of wellbeing because the inaccessibility of specific areas can restrict the capacity of individuals and social groups to participate in activities centralized in space. Lower subjective perceptions of the effects of transport on personal and collective liberties and empowerment are intimately linked with travel choices as to when to travel, where and with what means, central variables in transport planning and policy.

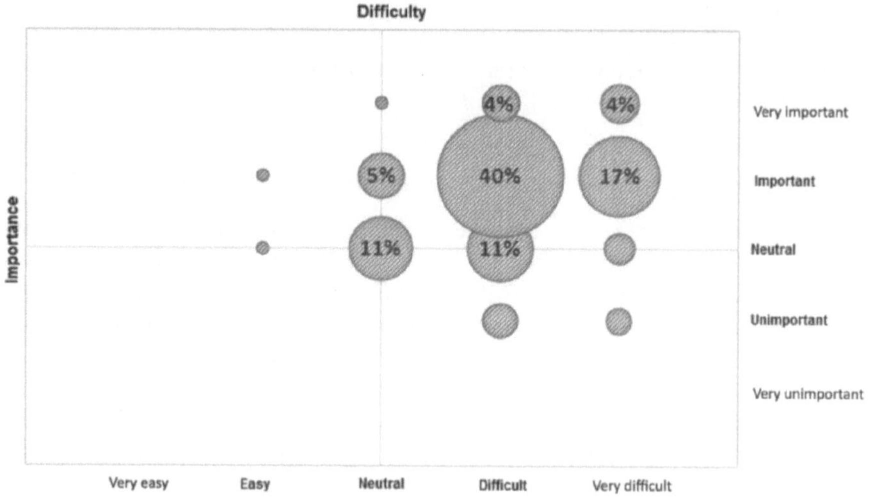

Figure 9.7 Ability to move at any time (% of respondents)

Source: Own elaboration based on field data

Respondents were asked about their top three priorities for public investment or actions to improve the transport system (Table 9.2). The results reveal a limited awareness of what the government can do to improve transport.

Conclusion

In a context such as Abuja's, marked by spatial and social differences embedded in a post-colonial modernist vision of development, and where issues of social diversity, spatial segregation and informality are often overlooked, a reflection on the wider implications of urban transport can contribute to inform urban transport development and policy. This is particularly true in relation to concerns often considered 'beyond' transport planning such as security, autonomy and freedom.

The evidence outlined here is unique in the Nigerian context, and not easily found in the literature concerned with transport and wellbeing in a Global South context. When considered together, the dimensions and sub-dimensions of wellbeing related to urban transport offer a deeper understanding of the dynamic relation between mobility and the wellbeing of residents and workers in Abuja. The framework provides a way to outline the evidence, allowing different interrelated variables to 'dialogue' with each other as in the case of access to opportunities and dominant socio-economic characteristics and perceptions. Indices are designed so they can be easily interpretable and interchangeable so the information can be accessible at most levels of urban transport development.

Respondents in Abuja's low-income satellite towns have very limited expectations of what the government can do to improve transport, highlighting infrastructure-related issues as their main priority. This contrasts with the results of our analysis of difficulties and importance of transport features, which highlight issues such as security, connectivity, local accessibility and autonomy as being more pressing. This is linked with the priorities of the planning system and the level of participation that it involves. Having a predominantly top-down approach, the plans for the city – outlined more than three decades ago – do not acknowledge the needs and preferences of the population arising from its considerable social, spatial and economic diversity.

Results suggest a mismatch between the actions and proposals by the state and the priorities for improving wellbeing of local citizens. On the one hand, the priorities revealed by both planners and decision-makers question the traditional transport system in practice in Nigeria. The ideological preference for the private vehicle and the strong focus on infrastructure and modernity do not respond equitably to the needs for improving wellbeing. In Nigeria, the issue of congestion is one of the most relevant by-products of such traditional systems, and it has a profound effect on the distributional impacts of daily mobility practices for different social groups. On the other hand, the study of wellbeing in its multiple dimensions calls for attention to issues frequently overlooked from mainstream transport planning. Furthermore, the ideological underpinnings of the wellbeing framework that supports fostering not only accessibility but the relational and subjective contributions of transport to people's lives contribute to debates around current approaches to transport development and decision-making and the role of mobility in human development.

Scaling up a methodology such as the one we propose, coupled with greater involvement of the local population in identifying indicators that build on their own interpretations of the contribution of transport to their wellbeing – considering the subjective beliefs and perceptions of people at different social positions – could go a long way in opening a more fluid and constructive interaction between the state and civil society. This can also offer a broader and deeper understanding of the nature of wellbeing as it relates to transport users than that offered by civil servants and transport experts. The formulation of plans is based on knowledge that, however sophisticated, can always be challenged and refined by the rich experience of people who use a system daily, and probably more regularly than those who make key decisions that can profoundly affect their lives.

Notes

1 This is part of the 'Urbanization Research Nigeria' (URN) Programme (2013–2017), by a consortium coordinated by ICF International (Bloch and Papachristodoulou, 2014). URN is funded by the UK government's Department for International Development (DFID). Quantitative data were collected in collaboration with a local consulting firm, STO Associates.

2 Identification of interviewees: male/female (M/F), age, occupation: civil servant (CS), academia (A), private sector (PS), municipal authority (MA), national government (NG)

References

Abdullahi, B.C., and Aziz, A.W. (2010). *Nigeria's Housing Policy and Public-Private Partnership (PPP) Strategy: Reflections in Achieving Home Ownership for Low-Income Group in Abuja, Nigeria.* 22nd International Housing Conference, ENHR, July, pp. 4–7.

Abubakar, I.R. (2014) Abuja city profile. *Cities* 41, 81–91.

Abubakar, I.R., and Doan, P.L. (2010). *New Towns in Africa: Modernity and/or Decentralization.* 53rd Annual Meeting, African Studies Association, San Francisco, USA.

Adeponle, B.J. (2013) The integrated city as a tool for sustainable development. *Journal of Educational and Social Research* 3 (5), 145–153.

Alkire, S. (2015). *Well-being, Happiness, and Public Policy. Centre for Bhutan Studies & GNH Research.* Thimphu, Bhutan

Amba, K. E. (2010). *Unmasking the Barriers to Citizen Participation in Political Processes in Abuja, Nigeria.* Report. University of Guelph. Ontario.

Australian Government (2010). *Australian Social Inclusion Board.* Annual Report, Commonwealth of Australia, Canberra.

Banister, D. and Bowling, A. (2004) Quality of life for the elderly: The transport dimension. *Transport Policy* 11 (2), 105–115.

Becker, G.S. (1962) Investment in human capital: A theoretical analysis. *The Journal of Political Economy* 70 (5) 9–49.

Beckman, L.J. (1976) Alcoholism problems and women: An overview. In: Greenblatt, M. and Schuck, M.A. (eds.), *Alcoholism Problems in Women and Children*, 65–69. New York, NY: Grune and Stratton.

Bloch, R. and Papachristodoulou, N. (2014) *Urbanisation Research Nigeria (URN) Framework and Plan.* London, UK: ICF International.

Boarini, R., Johansson, Å., and d'Ercole, M.M. (2006). *Alternative Measures of Wellbeing.* OECD. Paris

Camfield, L. and Skevington, S.M. (2008) On subjective wellbeing and quality of life. *Journal of Health Psychology* 13 (6), 764–775.

Cervero, R.B. (2013) Linking urban transport and land use in developing countries. *Journal of Transport and Land Use* 6 (1), 7–24.

Currie, G. and Delbosc, A. (2011) Exploring Transport Issues. In Currie, G. (ed.), *New Perspectives and Methods in Transport and Social Exclusion.* Emerald Bingley, UK.

Delbosc, A. and Currie, G. (2011) The spatial context of transport disadvantage, social exclusion and wellbeing. *Journal of Transport Geography* 19 (6), 1130–1137.

Dodge, R., Daly, A.P., Huyton, J. and Sanders, L.D. (2012) The challenge of defining wellbeing. *International Journal of Wellbeing* 2 (3), 222–235.

Ebo, I. (2015). *Planning for Exclusion in Abuja.* Available at: www.opendemocracy.net / opensecurity/ifeoma-ebo/planning-for-exclusion-in-abuja (accessed March 2016).

Falola, T. (1996). *Africa in Perspective" in Africa Now: People, Policies and Institutions.* Ed. S Ellis. London: Heinemman.

FCT MDG Office (2009) *FCT MDG baseline 2009: Employment. Federal Capital Territory Millennium Development Goals Office.* Available at: www.mdgfctabuja.net/Base line10/Employment.aspx (accessed June 2015).

Femi, S.A.G. (2012) Characterization of current transportation challenges in the federal capital territory, Nigeria. *Journal of Sustainable Development* 5 (12), 117.

Fowler, D, Huchzermeyer, M., Idahosa, J (2008) The Myth of the Abuja Master Plan: Forced Evictions as Urban Planning in Abuja, Nigeria. The Centre on Housing Rights and Evictions. Geneva, Switzerland). Home, R. K. (1983). Town planning, segregation and indirect rule in colonial Nigeria. *Third World Planning Review* 5 (2), 165.

Ilesanmi, I. (2006). *Pre-feasibility Assessment of Decentralised Sanitation Systems for New Satellite Settlements in Abuja*. Cuvillier. Abuja, Nigeria

IPA (International Planning Associates) (1979) *The Master Plan for Abuja, the New Federal Capital of Nigeria*. FCDA, Abuja: Nigeria.

Iro, I. (2007) Demographic pressure and the application of GIS in land reforms: The case of restoration of Abuja master plan and sanitization of cadastral and land registry. *In Proceedings of map middle east conference on GIS development*, Dubai, UAE.

Jibril, I.U. (2015). *Planning and Land Administration Challenges in Developing New Cities: The Abuja Experience in Nigeria*. FIG Working Week 2015: From the Wisdom of the Ages to the Challenges of the Modern World. Sofia, Bulgaria, 17–21 May.

Jones, P. and Lucas, K. (2012) The social consequences of transport decision-making: Clarifying concepts, synthesising knowledge and assessing implications. *Journal of Transport Geography* 21, 4–16.

La Placa, V., McNaught, A. and Knight, A. (2013) Discourse on wellbeing in research and practice. *International Journal of Wellbeing* 3 (1), 116–125.

Levy, C. (2013) Travel choice reframed: 'deep distribution' and gender in urban transport. *Environment and Urbanization* 25 (1), 1–17.

Lucas, K. (2012) Transport and social exclusion: Where are we now? *Transport Policy* 20, 105–113.

Mabogunje, A.L. (1968) Urban land use problems in Nigeria. *Institute of British Geographers Special Publication* 1, 203–218.

Mabogunje, A.L. (1990) Urban planning and the post-colonial state in Africa: A research overview. *African Studies Review* 33 (2), 121–203.

Mabogunje, A.L. (1992) New initiatives in urban planning and management in Nigeria. *Habitat International* 16 (2), 73–88.

Manderson, L. (2005) The social context of wellbeing. In: Manderson, L. (ed.), *Rethinking Wellbeing*. Netley, South Australia: Griffin Press.

McFarlane, C. (2008) Governing the contaminated city: Infrastructure and sanitation in colonial and post-colonial Bombay. *International Journal of Urban and Regional Research* 32 (2), 415–435.

Mollenkopf, H., Marcellini, F., Ruoppila, L., Flaschentrager, P., Gagliardi, C. and Spazzafumo, L. (1997) Outdoor mobility and social relationships of elderly people. *Archive of Gerontology and Geriatrics* 24, 295–310.

Mollenkopf, H., Baas, S., Marcellini, F., Oswald, F., Ruoppila, I., Szeman, Z., Tacken, M., Wahl, H. (2005) *Mobility and Quality of Life: Enhancing Mobility in Later Life*. ISO Press, Amsterdam.

Myers, G.A. (2011) *African Cities: Alternative Visions of Urban Theory and Practice*. London, UK: Zed Books.

The Nation. (2015). *Beyond Minimum Wage*. Available at: http://thenationonlineng.net/beyond-minimum-wage/ (accessed May 2015).

Office for National Statistics (2014). Measuring National Wellbeing, Life in the UK, 2014. *Database*. Available at: http://webarchive.nationalarchives.gov.uk/20160105160709/http://ons.gov.uk/ons/rel/wellbeing/measuring-national-wellbeing/life-in-the-uk – 2014/index.html (accessed 14 March 2015).

Odukogbe, S. (Photographer). (2015, March 20). Fieldwork photographs: Transport and wellbeing in Urban Nigeria [digital image compendium]. Private Archive. Development Planning Unit, University College London. London.

Ohnmacht, T., Maksim, H. and Bergman, M.M. (eds.). (2009) *Mobilities and Inequality*. Aldershot: Ashgate.

Ojo, V.O. (2011). *Customer satisfaction: A framework for assessing the service quality of urban water service providers in Abuja, Nigeria*. Doctoral dissertation. Ojo, VO.

Oviedo Hernandez, D., and Titheridge, H. (2015) Mobilities of the periphery: Informality, access and social exclusion in the urban fringe in Colombia. *Journal of Transport Geography* 55 (1), 152–164.

Pacione, M. (2003) Urban environmental quality and human wellbeing – a social geographical perspective. *Landscape and Urban Planning* 65 (1), 19–30.

Solanke, M.O. (2013) Challenges of urban transportation in Nigeria. *International Journal of Development and Sustainability* 2 (2), 891–901.

Spinney, J.E., Scott, D.M. and Newbold, K.B. (2009) Transport mobility benefits and quality of life: A time-use perspective of elderly Canadians. *Transport Policy* 16 (1), 1–11.

Stanley, J.K., Hensher, D.A., Stanley, J.R. and Vella-Brodrick, D. (2011) Mobility, social exclusion and wellbeing: Exploring the links. *Transportation Research Part A: Policy and Practice* 45 (8), 789–801.

Stanley, J., & Lucas, K. (2008). Social exclusion: what can public transport offer?. *Research in transportation economics* 22 (1), 36–40.

Stanley, J., & Vella-Brodrick, D. (2009). The usefulness of social exclusion to inform social policy in transport. *Transport Policy* 16 (3), 90–96.

Stiglitz, J., Sen, A., & Fitoussi, J. P. (2009). The measurement of economic performance and social progress revisited. Reflections and overview. Commission on the Measurement of Economic Performance and Social Progress, Paris.

Swanson, M. (1977) The sanitation syndrome: Bubonic plague and urban native policy in the Cape Colony, 1900–1909. *Journal of African History* 18, 387–410.

Titheridge, H., Christie, N., Mackett, R., Hernández, D.O., and Ye, R. (2014) Transport and Poverty. *A Review of the Evidence*. Report, University College London, London, UK. Available at: https://www.ucl.ac.uk/transport-institute/pdfs/transport-poverty (accessed May 2015).

Umoh, N.N.E. (2012) *Exploring the Enabling Approach to Housing Through the Abuja Mass Housing Scheme*. Doctoral dissertation, Massachusetts Institute of Technology, Cambridge, MA.

UNDP (1990). Human Development Report 1990. Oxford University Press, New York

UN-HABITAT (2013) *Planning and Design for Sustainable Urban Mobility*. Global Report on Human Settlements. Nairobi, Kenya.

van Lindert, P. (2016) Rethinking urban development in Latin America: A review of changing paradigms and policies. Habitat International, 54, 253-264.

Van Wee, B., & Geurs, K. (2011) Discussing equity and social exclusion in accessibility evaluations. *European Journal of Transport and Infrastructure Research* 11 (4).

Vella-Brodrick, D. (2011) Contemporary perspectives on wellbeing, In Currie, G. (ed.), *New Perspectives and Methods in Transport and Social Exclusion*. Emerald. Bingley, UK.

Vella-Brodrick, D. and Delbosc, A. (2011) Measuring wellbeing, In Currie, G. (ed.), *New Perspectives and Methods in Transport and Social Exclusion*. Emerald Bingley, UK.

Wachs, M. (2010) Transportation policy, poverty and sustainability. *TRR* 2163, 5–12.

White, S.C. (2010) Analysing wellbeing: A framework for development practice. *Development in Practice* 20 (2), 158–172.

10 Undertheorized mobilities

Fabiola Berdiel

Drawing on feminist and post-colonial perspectives on geography and urban sociology and scholars operating on the ground, this chapter proposes a conceptual framework built around the proposition that knowledge of undertheorized mobilities spaces in the Global South can deepen attempts to conceptualize and operationalize urban articulations and inform urban policies. By urban articulations we mean the ways urban residents organize to facilitate networks and flows of people, resources, spaces, discourses, and representations; seizing transitory opportunities and activating gaps in time and space as productive hubs within questionable economic and political infrastructures. By undertheorized mobilities spaces we mean heterogeneous configurations and dynamic, yet transitory, situations of socio-economic productivity and political creativity that are relatively invisible to current urban frameworks because they do not fit within existing analytical models. Binary theory models and policy frameworks, grounded in the experiences of cities in the West and North, ignore the diversity of geographies, practices, and possibilities 'in-between' dichotomous perceptions. The chapter argues that understanding these 'in-between' spaces is crucial to developing interventions that are appropriate to the transnational, differentiated, often syncretic character of our contemporary urban societies. The chapter draws on key propositions in emergent urban scholarship seeking to overcome the limitations of analytical categories and frameworks developed in the Global North, in an attempt to conceptualize, operationalize, and productively mobilize new and emerging urban articulations from the Global South. We argue that institutions must be capable of engaging with the complexities of the everyday practices of their constituents, developing innovative ways to conceptualize and materialize such engagements in order to reframe ways of working, and to formulate and implement innovative policies of socio-spatial inclusion which maximize the resourcefulness and optimize the mobilities of urban citizens and residents.

Introduction

> Mark Twain, once a steamboat captain on the Mississippi, developed techniques for navigating the river. While the passengers saw "pretty pictures" of landscape scenes, he was extracting information from the changing "face of the water."

> A little ripple, eddy, or "faint dimple" signaled turbulence or obstacles in a complex and potentially dangerous organization below the surface. These were markers of unfolding potentials or inherent agency in the river.
>
> Keller Easterling, Extrastatecraft, p. 18

The fields of urban inquiry are at a critical juncture. Much of our urban life and therefore, its supporting policies, are obscured by the binaries that shape our urban knowledge. Existing urban theory models and policy frameworks tend to be based on oppositions – juxtaposing extremes, such as developed and developing, western and nonwestern, modern and traditional, formal and informal – in which each paradigm is presumed to be unitary and homogeneous and only one can be dominant. This approach is limiting to explain the complexities of pressing global problems and the '*markers of unfolding potentials or inherent agency*' in the way urbanization currently unfolds. Binary theory models and policy frameworks, grounded in the experiences of cities in the West and North, ignore the diversity of geographies, practices, and possibilities 'in-between' these dichotomous perceptions. Further, the analytical focus of dominant theoretical frameworks fails to incorporate the collective, often cooperative knowledge produced inside the urban space itself, knowledge stemming from 'lived world' mobilities often informal and unregulated. The need to redirect urban theory, research, and practice away from old paradigms to respond to emerging realities of global urbanization such as environmental degradation, inequality in resource distribution, uneven development, and structural violence, is imminent. Urban inquiry approaches need to broaden their scope and epistemology to learn from the creative and dynamic agency of urban experiences in the Global South, collectively appropriating and reshaping urban spaces in an attempt to address these problems. This broadening implies a shift from a focus on recognizable infrastructure, policies, and institutions to an analytical approach that captures the innovative capacities of Global South urban residents to organize and collaborate despite functioning infrastructures, policy frameworks, and institutional practices in which to do so. In other words, shifting the focus from physical infrastructure to people as infrastructure, from government policies to informal arrangements, from formal institutions to social and religious networks.

Current critical urban scholarship challenges contemporary theoretical approaches, proposing new research agendas and lexicons, and revisiting 1970s debates surrounding the definition of the 'urban' (Brenner, 2013; Merrifield, 2013b; Pieterse and Simone, 2013; Roy, 2009). Although critical urban scholars approach such challenges from multiple positionalities – neo-Marxist, postcolonial, and feminist perspectives – the proposed research agendas coincide in bringing to light the multiplicity of economic, political, historical, ideological, technological, and social constructions that interact to produce urban spaces. This chapter seeks to address questions raised by current urban debates and to overcome the limitations of contemporary interpretative frameworks (Brenner, 2013; Merrifield, 2013b; Pieterse and Simone, 2013; Roy, 2009).

Drawing on key propositions in this body of literature, we propose a conceptual framework which makes the case that knowledge of *undertheorized mobility spaces* can deepen attempts to conceptualize, operationalize, and productively mobilize urban articulations and inform urban policies. By urban articulations we mean the ways urban residents organize to facilitate networks and flows of people, resources, spaces, discourses, and representations. These articulations seize transitory opportunities and activate gaps in time and space as productive hubs within questionable economic and political infrastructures. By *undertheorized mobility spaces* we mean heterogeneous configurations and dynamic, yet transitory, situations of socio-economic productivity and political creativity that are relatively invisible to current urban frameworks because they do not fit within existing analytical models. That is to say, the context and historically specific geographies, practices, and possibilities that characterize the ways urban residents organize to overcome the obstacles presented by unreliable institutions and limited infrastructure. To unearth the dynamics of these configurations this chapter builds on the theoretical concepts developed by scholars operating on the ground in Jakarta (AbdouMaliq Simone, 2013), Cape Town (Pieterse and Simone, 2013), Mumbai (Ananya Roy, 2011), Occupy (Andy Merrifield, 2013a), and other major global communities; triangulating these phenomena by situating themselves between the center and the edge, searching for agency.

We claim that if we (re)frame and productively engage specific sites of urban knowledge production in the Global South, we can inform better policymaking that takes a cross-disciplinary multi-level approach to: 1) build viable public institutions capable of engaging with the complexities of the everyday practices of their constituents; 2) develop innovative ways for public institutions to conceptualize and materialize such engagements in order to reframe their ways of working; and 3) formulate and implement innovative policies of socio-spatial inclusion which maximize the resourcefulness and mobilities of urban citizens and residents.

To reiterate, this chapter seeks to raise research questions and areas for further study that can enrich our research agendas beyond universalistic models that fail to account for context and historically specific human agency and local processes in the Global South. More specifically, the proposed framing aims to move beyond the mobility/immobility dichotomy and to identify the dynamic in-betweenness of movement, exchange, connection, and circulation in Global South cities.

Working definitions

> Uncertainty is seldom viewed as a resource, even though various aspects of uncertainty may be at the heart of contemporary urban growth.
>
> AbdouMaliq Simone, Cities of Uncertainty, p. 245

In this chapter, we borrow three of AbdouMaliq Simone's definitions to inquire into undertheorized urban mobility spaces: the 'in-between', 'productive uncertainty', and 'agency in times of uncertainty' (Simone and Rao, 2012; Simone, 2013). The notion of the 'in-between' is conceptualized and operationalized at two levels. First, 'in-betweenness' as a phenomenon aims to capture aggregation, alignment, stacking, and circulation of people, resources, spaces, discourses, and representations that seize transitory opportunities and activate gaps in time and space as 'points of reference, connection and anchorage' (Simone, 2013, 246). These heterogeneous configurations and dynamic, yet transitory, situations of socio-economic productivity and political creativity constitute what is referred to in this chapter as *undertheorized mobilities*. This leads us to the second notion of the 'in-between' as an epistemology: the way that we produce knowledge about these social, political, and economic formations, which lack clear narratives, distinctive class alignments, cohesive ideological coalitions, demarcated territorial identities, fixed collective interests and aspirations, and stable institutions that mediate and represent them. The apparent absence of these elements does not mean that these configurations do not exist; however, their invisibility in theoretical and policy frameworks often means that the needs of undertheorized mobility spaces go unrealized and unmet. Conversely, this chapter utilizes the notion of 'in-between' to focus our attention on the normality of everyday life – the life between the binary opposites, which is messy, fluid, and opportunistic – underexplored by mainstream analytical urban frameworks grounded in the experiences of the West and North. People, resources, spaces, discourses, and representations agglomerate, organize, mobilize, and circulate in this yet undetected space. The fact that these phenomena have been undertheorized by scholars and not operationalized by policymakers brings to light the limitations of existing theoretical and policy frameworks to capture these dynamic, yet diffuse, relational networks and practices in the Global South and post-colonial contexts (Simone and Rao, 2012).

Simone claims that uncertainty has become the norm rather than the exception (Simone, 2013: 2). Globalization unleashed productivity and wealth while simultaneously exacerbated inequality, marginality, and poverty. The world we live in today is characterized by fiscal crises, massive youth unemployment, environmental degradation, and vast migration. Tomorrow is no longer predictable without focusing our inquiries on multi-scalar dynamics and the creative and dynamic agency of context and historically specific human action and local processes. If we agree with Simone's claim that 'uncertainty is at the heart of urbanism, for urbanism is not a destination but a work always in progress' (Simone, 2013, 3), then focusing on agency and the in-between can provide useful theoretical and policy frameworks to understand geographies of possibilities through the actual practices of people navigating uncertainty at a specific time and space. From this perspective, in a world where uncertainty has become the norm,[1] we can begin to theorize and operationalize uncertainty as a resource rather than a challenge or limitation since it is under these circumstances where policy is played out and performed. How can policymaking capitalize on the aggregation and circulation of people, resources, spaces, discourses, and representations that circumvent structural constraints in the context of uncertainty to set up the conditions for new alignments and possibilities?

Finally, the notion of 'agency in times of uncertainty' is used to capture 'the capacity to construct elastic relationships with what is available to residents at any particular time and use these relationships as platforms to access new experiences and networks' as a base or framework (Simone, 2013: 2). For example, the way urban residents establish relational networks to take over land, connect small-scale economic operations, and circulate information to overcome dysfunctional systems of measurement, administration, and regulation. In unpredictable and unstable times, which is the normality of everyday life in most urban contexts – fiscal crises, social unrest, lack of political accountability and representation, and environmental degradation and climate change, to name a few – relational networks emerge to capitalize on the available resources, mainly social infrastructure, which is opportunistic and transitory rather than tied to established coalitions and networks.

In times of uncertainty when economic and political infrastructure is questionable – as it is in most Global South and post-colonial contexts – the emergence of social infrastructure presents a more nuanced understanding of the possibilities of 'agency in times of uncertainty'. This type of agency expands operational spaces to residents of limited means based on the arrangements they are able to devise with one another and the way they manipulate existing systems to adapt them to their own needs and aspirations. For example, in Cuba the partial opening of the socialist economy to small-scale entrepreneurship has allowed Cuban citizens to obtain licenses for a limited number of private business activities, among them renting out rooms to tourists. Obtaining a license for a private business implies paying extremely high taxes, which the majority of Cubans who work for the state do not pay. Through their relational networks, Cubans have established profitable operations for a large number of people by manipulating the existing system of licenses. One person obtains a license for renting out a room to tourists. Once tourists arrive, they are offered a wide range of services, including food, cigars, money exchange, tourist guides, escorts, and drivers. Taxes are only paid for the license to rent a room for which all of the beneficiaries contribute their share. In this scenario, agency in times of uncertainty allows everyday citizens to circumvent systems of measurement, administration, and regulation and find affordable ways to fulfill their aspirations to profit from the opening of the economy.

Conceptual framework

Our interest in undertheorized urban mobility spaces of 'in-betweenness' and 'agency in times of uncertainty' is deeply influenced by the path-breaking work of Henri Lefebvre and Michel de Certeau. Both Lefebvre and de Certeau develop their work simultaneously with Michel Foucault's work on spatiality and power, to explore how social relations are produced and reproduced through overlapping dimensions that are simultaneously material, ideal, and quotidian. Lefebvre and de Certeau situate their work in Foucault's heterotopic spaces,[2] which highlight the complex dynamics among collective modes of administration, individual modes of reappropriation, and spaces of deployment. Both scholars challenge

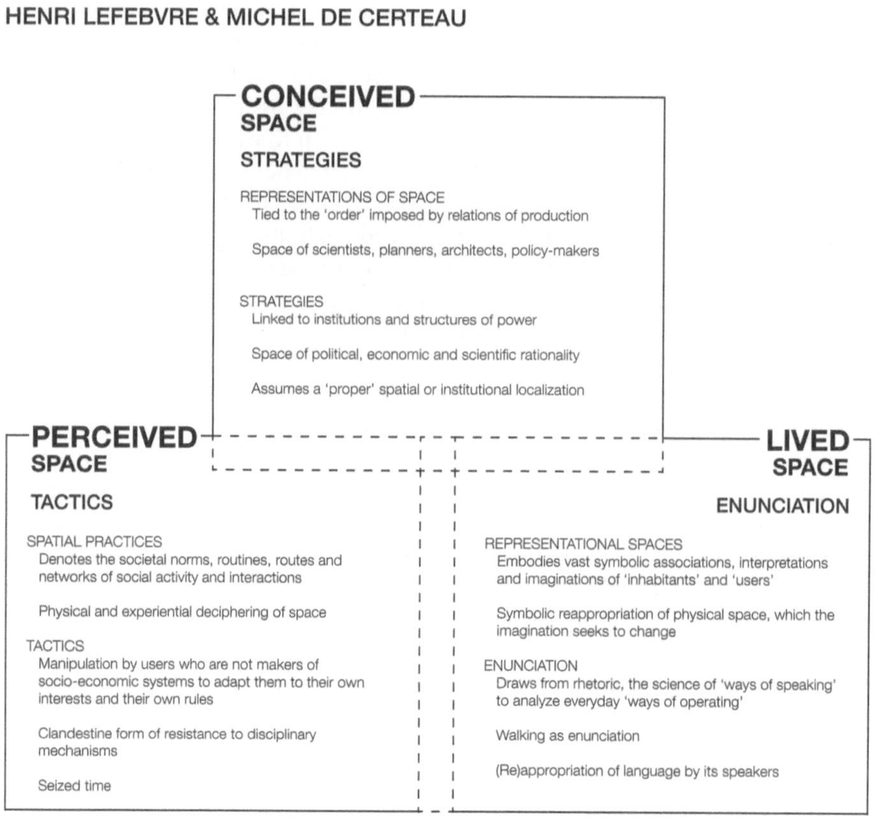

Figure 10.1 Production of space and practice of everyday life

binary theories of the urban, which limit our understanding to oppositions and contrasts, by introducing unitary theories that capture the physical, mental, and social spaces of material engagement, scientific conception, and societal expression (Figure 10.1).

In *The Production of Space (1974*) Henri Lefebvre proposes a 'conceptual triad' to capture the way urban spaces are incessantly being constructed at the intersection of 'representations of space' by architects, planners, and developers (conceived space); 'representational spaces' embodying vast symbolic associations, interpretations, and imaginations of 'inhabitants' and 'users' (lived space); and 'spatial practice' denoting the routines, routes, and networks of social activity and interactions (perceived space). Conceived space is the imagined space of representations tied to the 'order' imposed by relations of production and power. Lived space is the symbolic reappropriation of physical space, which the imagination seeks to change. Finally, perceived space is the real material space of locality, the way that we physically and experientially decipher space. 'Representations of space', 'representational spaces', and 'spatial practice' overlap and incorporate elements of one another to compose a unionized theoretical structure of 'the actual production of space by bringing the various kinds of space and the modalities of their genesis together' (Lefebvre, 1991 [1974]: 16).

In his work, *The Practice of Everyday Life (1984)* de Certeau brings to light the everyday practices by which users reappropriate the space conceived by institutions and structures of sociocultural power. To explore this proposition further, de Certeau introduces the notion of 'strategies' and 'tactics'. Strategies can be situated in Lefebvre's conceived space. 'Strategies' are linked to institutions and structures of power and are produced in the space of political, economic, and scientific rationality. 'Tactics' mainly inhabit in Lefebvre's perceived space. 'Tactics' are a clandestine form of resistance to disciplinary mechanisms exercised by taking advantage of timely opportunities that must be seized on the spur of the moment. To this juxtaposition of two forms of agency proposed by de Certeau and produced in Lefebvre's conceived and perceived spaces, we incorporate a third form of agency to complete a conceptual triad, which exists in Lefebvre's lived space and is proposed by de Certeau as the (re)appropriation of space by its users through 'enunciation'. Walking as enunciation is users' way of reappropriating networks of discipline, topological systems, contractual relations, and resources through movement and interpretation by creating their own trajectories and imaginaries. Similar to Lefebvre's proposition of the production of social space through the continuous interaction and collusion of conceived, perceived, and lived dimensions of space, de Certeau argues that the urban is generated by the strategies of institutions of power, the tactics of walkers in the city, and by the practices of everyday life in the form of enunciation (de Certeau, 1984). For these practices to overcome the blindness of theory (strategies) and daily experience (tactics), the notion of walking as enunciation as an epistemology focuses our attention on the interplay of all three spheres in the social production of urban spaces of 'in-betweenness' and 'agency in times of uncertainty'

In the broadest sense, we are interested in scrutinizing if an inquiry into the phenomenon of 'in-betweenness' and an exploration of the unfolding potentials of 'agency in times of uncertainty' can overcome the limitations of existent binary urban research and policy frameworks and deepen the possibilities inherent in the way people organize, mobilize, and operationalize available resources in times of uncertainty in the Global South.

With this in mind, we build on Lefebvre's and de Certeau's epistemologies, to foreground a conceptual triad that breaks away from binary understandings; serves as a tool to scrutinize if 'in-betweenness' exists as a distinct phenomenon that is conceived, perceived, and lived, and that can be observed and theorized; and accounts for 'agency in times of uncertainty'. The conceptual notions that we propose inquire into *'in-between' geographies*, which break away from Lefebvre's conceived space and de Certeau's strategic power; *'in-between' practices*, which are produced within Lefebvre's perceived space and de Certeau's tactics of resistance; *and 'in-between' possibilities*, which come to life in Lefebvre's lived space and are shaped through de Certeau's walking as enunciation of possibilities. These conceptual notions are meant to be used as new entryways, new positionalities, to focus attention on a dynamic process of interconnection, interplay and mobilities taking place in the gaps between clearly denoted urban institutions, spaces, and actions.

Our claim is not that these concepts are new. Rather, we are interested in exploring how urban scholars, proposing a redefinition of the urban agenda, are applying and actualizing Lefebvre's and de Certeau's ideas to the framing of policy-relevant research. By situating our inquiry in these spaces, we can unearth the 'in-between' obscured in binary constructions as well as deepen our understanding of binary terms.

The proposed conceptual framework seeks to contribute to an agenda for twenty-first century urban inquiry that focuses on a variety of dynamic under-theorized mobility spaces, superimposing the notions of 'in-between' geographies (conceived/strategies), practices (perceived/tactics), and possibilities (lived/enunciation) on Lefebvre's and de Certeau's unitary theories.

In-between geographies

> Forms of worlding cannot be understood merely as a globalization imposed by the West on the Rest.
>
> Ananya Roy, 2014

The notion of *in-between geographies* is used to capture Lefebvre's conceived space and de Certeau's strategies interpreted as Simone's, Pieterse's, and Roy's layered urban spaces with variegated histories, geographies, and mobility experiences that do not conform to binary urban theory models and policy frameworks (Simone, 2001, 2004a, 2004b, 2012, 2013; Pieterse and Simone, 2013; Roy, 2009, 2011, 2014). Overall, the notion of *in-between geographies* relocates the 'urban center' to multiple cores and peripheries that through movement, encounter, and exchange facilitate complex circuits of support; highlights the multiple temporalities and governance systems that overlap and exist simultaneously in space; and explores the differential deployment of state power dependent on a hierarchical conceptualization of space.

Multiple cores and peripheries

Particularly important for expanding our scope of *in-between geographies* of theory is Ananya Roy's work on the 'worlding of cities' (Roy, 2009) that shifts the focus from a hierarchical analysis of cities based on economic competitiveness to a theorization of multiple cores and peripheries that 'cannot be understood merely as a globalization imposed by the West on the Rest' (Roy, 2014: 10). Roy argues that the majority of urban theory has been conceptualized and situated in the experience of Western Europe and North America, developing a canon and policy-framing grounded in the experiences of 'global cities' – those cities whose agglomeration economies positioned them as central nodes for the operation of the global system of finance and trade: Chicago, New York, Paris, Los Angeles, London. The notion of *in-between geographies* aims to, firstly, challenge the focus on economic competitiveness because it has the propensity to omit the terrain of politics and agency through which space is inhabited and contested. Secondly, position the theoretical frameworks developed in the Global South as heuristic devices that can be deployed to study '*all* cities from *this* particular place on the map' (Roy, 2011: 822, original emphasis) to overcome the binary created by the traditional canon, juxtaposing 'global cities' (Western Europe and the United States) to 'mega-cities' (Global South). Finally, focus our attention to geographies of connections that exceed their local origins through global mobility and networks, challenging the dominant focus on *located* global cities and their 'trait geographies', which neglect the richness of *dis-located* 'process geographies' of transnational flows such as those of trade, travel, and labor (Roy, 2009: 822).

To address the limitations of the current scope of geographies of theory, *in-between geographies* borrows Roy's notion of *multiple cores* to reposition Global South knowledges as heuristic devices that use strategic essentialisms[3] to deepen our conceptual frameworks for theorizing about *the urban*. Roy raises important questions to consider: How can Latin American frameworks on insurgent citizenship; South Asian frameworks on post-colonial forms of agency and subjectivity; East Asian frameworks on the polytemporal and polyvalent productions of globalization and global cosmopolitanism; Middle Eastern frameworks on the city as a site of performance for nation-building processes; and African frameworks on the reproduction of the city through circulations and transactions of social relations enhance the research and policy agenda for *all* cities (Roy, 2009)?

Adding a second dimension to *in-between geographies*, the notion of *multiple peripheries* draws on Abdoumaliq Simone's extensive work on the importance of the periphery in urban life to foreground how new *in-between geographies* of theory are created through 'movement' and networks of 'elastic security', concepts explained later in the chapter. For our analytical purposes, peripheries are defined as twofold: cities that have been 'at the periphery of urban analysis' (Simone, 2010: 14) and 'space in-between . . . never really brought fully under the auspices of the logic and development trajectories that characterize a center' (Simone, 2010: 40). *In-between geographies* borrow Simone's conceptualization of peripheries as generative spaces of possibilities, as another heuristic device to destabilize the center.

The potential of the concept of *multiple peripheries* lies in transcending localized geographies and incorporating users' agency in this transgression of bounded spaces. To further expand on this heuristic device, it is imminent to incorporate the notion of 'movement' to capture the expansion of livelihood opportunities through mobility and the respatialization of social institutions that cross politically demarcated geographical boundaries. In *For the City Yet to Come*, Simone illustrates the notion of 'movement' through a case study of Africans working in Saudi Arabia. Simone explores how a Sufi Muslim social institution is transformed into an organizing and unifying body, facilitating collaboration among diverse groups by reviving shared historical narratives. Sufi Muslim organizations have historically been a site of organizing; however, the nuance of Simone's analysis is how this relational network was operationalized outside of Africa to extend connections and expand opportunities for Africans outside of the continent.

Complementary to 'movement' is Simone's exploration of 'elastic security' in *Cities of Uncertainty*, which examines the power of information sharing through the performance of ethnic, regional, and religious identities to secure access to opportunities. *In-between geographies* builds on the notion of 'elastic security' to capture the complex circuits of support, based on reciprocal favors and opportunistic connections, facilitated by 'wearing multiple hats'. The idea of 'wearing multiple hats' alludes to the fluid performance of ethnic, regional, and religious identities at opportunistic times to broaden social networks for scaling-up opportunities and getting access to information. Visibility is at the core of 'elastic security', allowing others to see the potential of collaboration. For example, residents in Jakarta might attend mosque services, local mosques being the most prolific of local institutions, not necessarily out of religious devotion but to 'use the obligations of zakat – alms giving – as a mechanism of redistribution and thus a way of maintaining relationships between richer and poorer residents' (Simone, 2013: 252). Beyond zakat, being seen at the mosque facilitates reciprocity in other aspects of local life.

Mobility, visibility, and information sharing are key to function in the developing world where institutions are at times unreliable, infrastructure is limited, and alternative strategies are a crucial element for navigating urban spaces in this context. These new *peripheries* activated through mobility, visibility, and information sharing serve as an instrument to overcome regulatory frameworks and institutions, highlighting the importance of agency in creating generative spaces of possibilities beyond geographical locality. It is important to highlight that these strategies of mobility, visibility, and information-sharing transcend geographical locality because actors carry them to any context in which they operate, whether the developing or developed world. Understanding these strategies takes in-between geographies to another level, since these strategies are embedded in human agency rather than a specific location in the world.

Finally, the emergence of *core-periphery structures* as a component of new *in-between geographies* of theory allows us to explore global circuits of work that link peripheries to cores (i.e. the Philippines and Indonesia to Hong Kong, Central and South America to the United States) and trajectories of migration and relocation that link peripheries to peripheries (sub-Saharan Africa to India, Saudi Arabia, Thailand, Malaysia). The promise in these *core-periphery structures* is that they challenge conventional core-periphery dichotomies and allow us to capture 'worlding from below', the agency of users taking advantage of the urban as a site of encounter and exchange to reach out and operate at the level of the world.

Multiple temporalities and governance systems

In addition to challenging the territorial localization of in-betweenness, the notion of *in-between geographies* seeks to capture the multidimensionality of time and governance structures. In *Rogue Urbanism*, Edgar Pieterse introduces the concepts of 'Palimpsests' and 'Governmentalities' to elucidate 'the specificity of everyday life [urbanism] and its underlying material, ecological and symbolic currents' (Pieterse and Simone, 2013: 13). We build on these two propositions to provide a framework for theorizing and operationalizing two notions, particularly relevant for post-colonial contexts where pre- and post-colonial histories and governance structures overlap and coexist: *multiple temporalities* and *multiple governance systems. Multiple temporalities* capture Pieterse's 'palimpsests' by bringing our attention to the overlapping historical narratives, social norms, symbolic economies, and imaginaries that function simultaneously in coexisting and co-constructing temporalities. 'Palimpsests' allow us to understand historical traces and the flow and overlap of time and spatial imaginaries. For example, Cuba today is experiencing two seemingly contradictory imaginaries, a political model with foundations in the values of the 1959 Revolution (equality and social justice) and market-driven policies and practices which prioritize economic performance. Reducing this historical moment to a linear interpretation of a transition into capitalism ignores the complex dynamics of navigating an ideological duality, which is prevalent in institutions and formal practices as well as in informal relations and everyday life. Although Cuba is by no means a capitalist society, they are experiencing the commodification of social relations by the illusion of the market that is yet to come. The state is commodifying the socialist dream to attract foreign investment and 'performing' a socialist discourse while implementing market-driven policies. The citizenry seeking to fill the void in legitimacy, ideology, and structure left by the disconnection between a socialist discourse and market-driven practices is exerting their presence and participation in the public sphere and through interstitial practices, opening spaces of negotiation, alternative 'publics', economies, and plausible futures. It is by inquiring into the multiplicity of historical narratives and symbolic economies that we are able to explain an otherwise unintelligible dynamic.

Similarly, the notion of *multiple governance systems*, derived from Pieterse's 'governmentalities', seeks to capture grounded, historicized, and situated accounts of power dynamics by focusing on contestation, adaptation, and re-legitimization of institutional power. This concept brings to light new social arrangements that circumvent the neoliberal governance apparatus, confronting it through mobility and everyday practices. These new social orders diffuse and fragment the traditional notion of state power by forcing us to look from the ground up at the *multiple governance systems* that shape urban space, which is particularly important for mobilities in developing contexts given the contestation of power structures taking place at multiple levels. An example of these multiple governance systems is Caroline Wanjiku Kihato's research in *Rogue Urbanism* on the complex dynamics between police raids of markets in Johannesburg looking for foreign migrants and the 'warning and representation system' developed by foreign migrants in which they alert one another of police presence and hire a South African spokesperson to represent them in negotiations with the police (Pieterse and Simone, 2013: 325). Urban governance in this instance lies at the intersection of state rules and regulations and more localized forms of discipline and regulation.

Multiple hierarchies in deployment of state power

Completing the conceptual notion of *in-between geographies* that proposes a redefinition of core and periphery and an expansion of our flattened understanding of space and time, *in-between geographies* incorporates an analysis of the way power is deployed and negotiated in various spatialities. In *Slumdog Cities*, Roy introduces the concepts of 'zones of exception' and 'gray spaces' to bring to light a 'system of graduated zones' that are either valorized or criminalized through the deployment of state power. 'Zones of exception' range from transnational zones of investment and cheap labor to transnational zones of refugee administration. The differential treatment of zones of exception and the contrasting deployment of state power in these geographies provides a useful critical lens to examine and challenge how the administration of space is operationalized according to hierarchies of value (economic vs. political). 'Gray Spaces' build on the notion of 'zones of exception' and differential deployment of power to denote spaces that exist between 'legality/approval/safety' and 'eviction/destruction/death'. Gray spaces are the 'periphery of peripheries', where 'bare life' is the norm rather than the exception. For example, the marginalization and displacement of Israeli Bedouins and Israeli Arabs by an ethnocratic Israeli state is illustrated in the work of Oren Yiftachel (2009). These spaces capture the way that the state 'launders' gray spaces created from above by powerful or favorable interests' and 'solves the problem of marginalized gray spaces through destruction, expulsion or elimination' (Roy, 2009: 235). Having a framework to understand the differential deployment of power in hierarchical spaces is crucial for a new theorization of *in-between geographies*.

In-between practices

> *Ruins not only mask but also constitute a highly urbanized social infrastructure. This infrastructure is capable of facilitating the intersection of socialities so that expanded spaces of economic and cultural operation become available to residents of limited means.*
>
> *AbdouMaliq Simone, 2004a*

In-between practices refers to Lefebvre's perceived space and de Certeau's tactics elucidated in Simone's and Pieterse's work exploring the ways 'inhabitants' and 'users' develop modes of operating in urban geographies in efforts to circumvent technologies of power and mobilize what is available to fulfill their aspirations.

This is particularly important to mobilities in the Global South because they are primarily informal and unregulated as the result of an immediate need to overcome the obstacles presented by unreliable institutions and limited infrastructure and resources. The notion of *in-between practices* seeks to explain courses of action that otherwise appear simply as diffuse and illogical dimensions of urban life (de Certeau, 1984; Simone, 2001, 2004a, 2004b, 2013; Simone and Rao, 2012). Overall, the notion of *in-between practices* seeks to capture modes of production of space that are unbounded by institutional and physical space and where collective agency takes advantage of moments and mobility opportunities to fulfill their needs outside of regulatory frameworks.

As perceptively noted by Pieterse and Simone (2013), distinguishing between the 'formal' and 'informal' is 'virtually meaningless across most [Global South] cities . . . because most of the city is the consequence of hybrid economic practices' (Pieterse and Simone, 2013: 14). Both Pieterse and Simone make the case that focusing on emerging flows and networks and *deal-making* can contribute to a better understanding of the potentiality of the Global South and post-colonial contexts, allowing researchers to foreground highly dynamic practices such as inventive entrepreneurship.

Particularly important for expanding our understanding of urban practices to incorporate collective clandestine agency is Abdoumaliq Simone's work on 'people as infrastructure' (Simone, 2004a) that extends the modes of provisioning, connecting, and assembling the city, usually associated with physical infrastructure, to people's activities. The notion of *people as infrastructure* focuses our attention on the economic and cultural collaboration among residents, seemingly at the margins of urban life, 'which is capable of generating social compositions across a range of singular capacities and needs (both enacted and virtual) and which attempts to derive maximal outcomes from a minimal set of elements' (Simone, 2004a: 410). In his case study, Simone examines the ways in which economic and social collaboration in everyday practices enable the diverse immigrant population of Johannesburg to negotiate and navigate the inner city, circumventing xenophobia and ethnic enclaves by building on translocal and multilateral transactions.

> [Common] national identity can provide a concrete framework for support among individuals who may have very different kinds of jobs. . . . These articulations are used by larger corporate groupings – cutting across several national identities – that facilitate various business efforts through subcontracting arrangements. One such enterprise might draw on the professional legitimacy of teachers, use their students as potential customers or corporate informants, and incorporate the trading circuits developed by petty traders and the repair skills of mechanics.
>
> (Simone, 2004a: 416).

The notion of *people as infrastructure* highlights how social infrastructure enables residents of limited means to expand their access to opportunities by facilitating the circulation of social networks, which become a platform for deal-making, modes of production, and institutional forms. Such an assemblage is highly mobile, provisional, and constantly negotiated because it draws its regularity on a process of constant adaptation to an unexpected range of scenarios, dependent on 'the particular histories, understandings, networks, styles, and inclinations of the actors involved' (Simone, 2004a, 410).

Simone's notion of *people as infrastructure* and Pieterse's *deal-making*, which are central to a theorization of *in-between practices*, extend Lefebvre's 'representations of space', linking spaces to specific identities, functions, and lifestyles to make them legible for specific actors at specific places and times, as well as linking modes of organization to overlapping scales. Simultaneously, this notion draws from de Certeau's tactics to capture the opportunistic utilization of socio-economic and cultural systems by scrutinizing the way that urban residents operate at the interstices of strategic limitations and (mis)represent their multiple identities, skills, and needs in order to function productively in questionable political, economic, and institutional infrastructures.

The critical question raised by incorporating *people as infrastructure* and *deal-making* – propositions which challenge the dichotomy of 'formal' and 'informal' – in our conceptual framework is how researchers, policymakers, and activist scholars can design ways of theorizing and operationalizing urban practices that are characterized simultaneously by regularity and provisionality. Visibility facilitates analysis and planning; however, the emphasis of existing frameworks has been on situating and defining the urban environment, populations, flows, structures, and activities to codify them according to recognizable and familiar systems, grounded in the experiences of cities in the West and North. The *in-between practices* captured by Simone and Pieterse are currently invisible to these frameworks because they are constituted through the ability of individuals to move across and between a wide range of spatial, cultural, economic, institutional, and transactional positionalities (Simone, 2004a: 408). It is this invisibility which often renders the needs of undertheorized mobility spaces to go unrealized and unmet.

The proposed framework seeks to reconceptualize the codification of belonging and mobilities beyond a logic of static identities and territorial representations to incorporate shifting identities and provisional networks of social collaboration that generate possibilities through their disarticulation of coherent urban space. As illustrated by the small-entrepreneurship Cuban schemes discussed in the 'Working definitions' section, multiple hats facilitate mobility: history professors become taxi drivers and cultural guides; escorts become business partners for foreigners wanting to invest; airbnb hosts transform into tourist agencies and currency exchange centers and nothing is what it appears to be but so much more.

This second component of the proposed conceptual framework incorporates an understanding of urban geographies emerging from provisional practices embedded in a wide range of tactics that residents use to circumvent limitations and surveillance, and to increase access to information, opportunities, and networks of collaboration. An understanding and theorization of this transgression requires a shift from a focus on formal infrastructure, policies, and institutions to an analytical approach that captures 'varied capacities of diverse urban residents to operate in concert without discernible infrastructures, policy frameworks and institutional practices in which to do so' (Simone, 2004b: 13).

In-between possibilities

> [The politics of encounter express] an encounter between people that has become an encounter between citizens, people who no longer ask for their rights, for the rights of man, for the right to the city, for human rights: these citizens meeting one another make no rights claims, posit no empty signifiers. They don't even speak – not in the conventional sense of the term; they just do, just act, affirm themselves as a group, as a collectivity, as a 'general assembly', wanting to take back that which has been dispossessed.
>
> Andy Merrifield, 2013a

Completing the three-dimensional approach, *in-between possibilities* is used to foreground Lefebvre's lived space and de Certeau's 'walking as enunciation of possibilities' interpreted by Merrifield's urban spaces of resistance and re-appropriation as the contemporary agora leading to socio-economic transformation (Merrifield, 2013a, 2013b). Further, this last dimension of *in-between possibilities* brings our inquiry back to its theoretical grounding by returning to Lefebvre's epistemological proposition of *transduction*, which reframes urban inquiry as a transdisciplinary endeavor, which unearths the 'possible object' of urban society as a horizon rather than an accomplished and concrete social reality located in actual time.

Politics of encounter

Andy Merrifield's exploration of the twenty-first century 'new urban question' and 'the politics of encounter' are of central importance for scrutinizing *the possible* in terms of place, politics, and the everyday. In his scholarly inquiry on the twenty-first century urban question, Andy Merrifield expands on Lefebvre's legacy by writing about *the possible* as Lefebvre would have if he were alive today. Merrifield repositions the urban as a place, a site for action and politics, rather than an actor itself. This proposition emphasizes Lefebvre's reframing of the urban as urban society instead of urbanization as the object of inquiry. Building on Lefebvre's 'signs of the urban' as 'signs of assembly' (Lefebvre, 2003: 118), Merrifield highlights that the role of the urban is as a socio-spatial sphere in which the 'betweenness' of people is intensified to capture popular imagination and extend possibilities beyond locality. Merrifield conceptualizes possibilities beyond Lefebvre's proposition of 'The Right to the City', arguing that this concept emerged from the 1960s urban struggles, where the right to the city took shape in a physical sense, the right to access the centrality of urban life and transform the city. Today, Merrifield argues, the city and revolutionary politics are no longer the same. Centrality in the contemporary global urban society should not be perceived as 'being at the center of things' geographically but as shifting centers activated by social acts of mobilization and negotiation, facilitated by human mobilities and the encounter in-between citizens that produce social space and define the urban.

With this in mind, we borrow Merrifield's notion of 'the politics of the encounter' as a new framework of agency to scrutinize the urban, politics, and the everyday. In the politics of encounter, Merrifield blends Lefebvre's 'production of space' and de Certeau's 'elucidation' to capture how urban encounters 'comprise the construction of use values as opposed to the appropriation of exchange values' in urban politics (Merrifield, 2013a: 916). The notion of urban encounters connects the local to the global, highlighting the impact of world economic and political trends on local practices and signaling to users (rather than consumers) their opportunities and limitations within 'the fragile planetary ecology [and a] rapacious global economy' (Merrifield, 2013b: 916). The notion of urban encounters repositions residents as users with agency as opposed to passive consumers of exchange values, overcoming the obstacles that promote passiveness of relationships and creating room for possibilities.

Merrifield's work incorporates an understanding of *the possible*, emerging from the urban as a place of assemblage that brings the *betweenness* of people, resources, spaces, discourses, and representations together. This coming together transforms their possibilities and consolidates their agency. An understanding and theorization of *in-between possibilities* requires shifting the conversation beyond a geographically fixed right-based agenda to focus our attention on a more open and horizontal appropriation of urban space by its users to 'affirm themselves as a group, as a collectivity, . . . wanting to take back that which has been dispossessed' through the encounter urban spaces facilitate (Merrifield, 2013b: 918). The recent upheavals in Tunisia, Egypt, Greece, and Spain and the Occupy Movement are ideal illustrations of the performance of *the politics of encounter*.

Transduction

The *possible* as a theoretical concept has been underexplored because of scientific scholarship dominated by a focus on what can be proven by empirical evidence and theorized based on previous frameworks. In his provocative work *The Urban Revolution*, Lefebvre challenged the 'scientific' canon by proposing to build a theory from an anticipatory hypothesis instead of building it on fact-based empiricism to critique urbanism as an ideology and institution, arguing that 'it establishes a repressive space that is represented as objective, scientific, neutral' (Lefebvre, 2003: 181). *In-between possibilities* builds on Lefebvre's epistemology of *transduction* to reframe urban inquiry as a transdisciplinary endeavor, which unearths the 'possible object' of urban society in its inquiry. In this conceptual framework, the possibilities of urban society are defined by a direction, an anticipatory hypothesis that elucidates social reality and simultaneously activates the future. *In-between possibilities* as an epistemological tool seeks to build on Lefebvre's understanding that to realize urban society we must confront the obstacles that currently render it impossible by developing tentative concepts (meaning); contextualizing them in historical processes and social realities (content); discovering their limitations in dialectical interaction with empirical research (boundaries); and continuously redefining concepts, advancing theory, and guiding praxis as part of an immersive engagement with real-time developments in society (possibilities). In the proposed conceptual framework, urban inquiry should be conducted by developing strategic concepts that simultaneously grasp historical and social processes, critique the present condition, and allow prospective co-construction of the future.

Planning policy implications

This chapter argues that it is time to mobilize and instrumentalize knowledge produced in *'in-between' geographies* towards new opportunities, *practices*, and *possibilities* that have been undertheorized in binary urban policy discourses. The dichotomous understanding of the urban limits our understanding of the dynamism, interplay, and mobilities of emerging realities of global urbanization and the collective power of differential urban experiences in the Global South and post-colonial contexts. More specifically, this chapter poses the following policy questions:

- How can a new urban agenda reframe the tools and possibilities of urban policymaking (infrastructure, economy, governance, inhabitants' agency) to 'in-between' geographies, practices, and possibilities for successfully coping with uncertainty?
- How can current models, lexicon, and epistemology of urban research be improved to provide better guidelines for policymaking?
- How can a cross-disciplinary multi-level approach to urban inquiry inform better policymaking to build viable public institutions capable of engaging with the complexities of the everyday practices of their constituents; develop innovative ways for public institutions to conceptualize and materialize such engagements to reframe their ways of working; and formulate and implement innovative policies of socio-spatial inclusion which maximize the resourcefulness and mobilities of urban citizens and residents?

Policy framing that truly operates within the proposed three-dimensionality must simultaneously: 1) examine the representations of space and strategies, that is, the space of scientific, economic, and political rationality; 2) analyze spatial practice and tactics, the physical and experiential processes related to users' deciphering and adapting of space; and 3) integrate into the analysis the spaces of representation, the symbolic re-appropriation of physical space, and the ways of operating in lived experience.

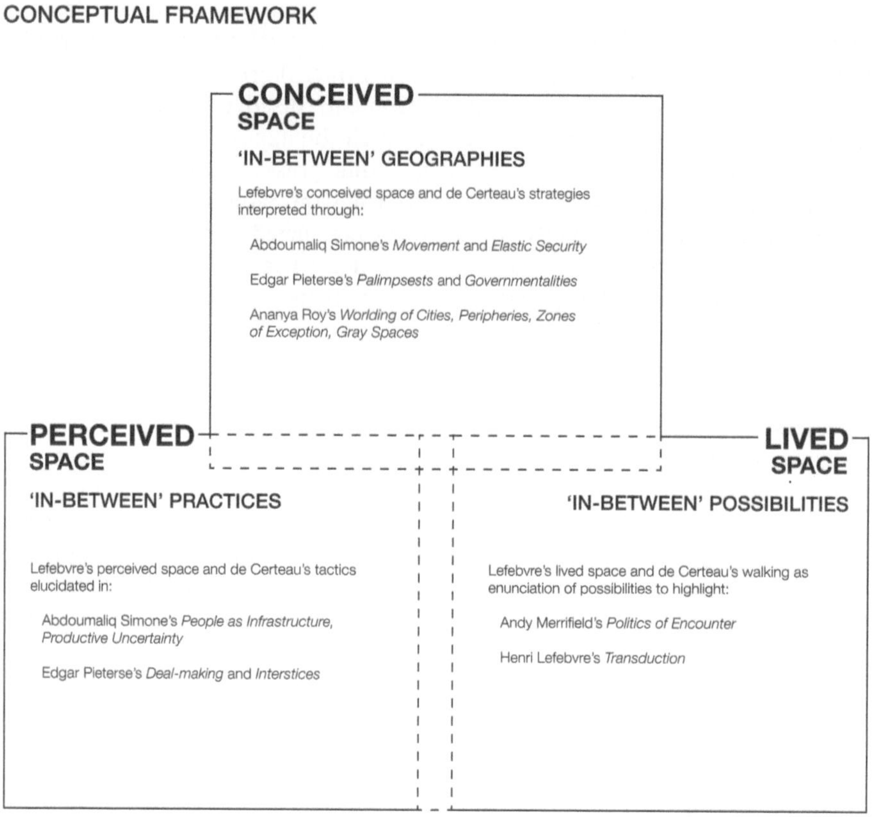

Figure 10.2 Conceptual framework for 'in-betweenness'

Conclusion

What is at stake in this chapter's proposition is a broadening of the scope, epistemology, and analyses of geographies, ways of operating, and potential agency inherent in the heterogeneity and multiplicity of urban articulations in the Global South. In the proposed conceptual framework, agency, action, and encounters are at the center while the urban is not an actor in itself but a place of in-betweenness where the possible can be mobilized. We cannot identify phenomena without giving it a name. Consequently, we are unable to act on that phenomena without proposing interpretations of the dynamics at play in that space. To communicate these explanations, we require terminology, conceptual notions, images, and diagrams that delimit and denote this space. Thus, proposing a different framing and naming for the location and mode of production of knowledge about *the urban* to inform better policymaking in the Global South constitutes an intervention in the debate.

This chapter's contribution to the debate on the redefinition of the current urban agenda is twofold. As an intervention in the epistemologies and methodologies of urban inquiry, we draw from the conceptual frameworks of the proposed research agendas to develop a theoretical framework that is able to capture and unearth the phenomenon of 'in-betweenness' that manifests itself in a diversity of localities, spaces, and geographies. As an intervention in urban activist scholarship, we seek to reframe the conversation to focus on the notion of 'agency in times of uncertainty', the way people make choices and act in the world around them that cannot be captured by frameworks such as 'citizenship', 'governance structures', or 'institutions' because it is heterogenous, contradictory, temporary, and performative.

Notes

1 This claim can be illustrated by this very narrow list of examples between 2011 and 2015: The economic collapse in Greece and Spain; the structural violence and social unrest in the United States, Mexico, Egypt, Brazil, and South Africa; the earthquake in Nepal, tsunami in Japan, hurricanes in New York and New Orleans, among many other natural disasters; the Ebola outbreak in Africa, avian flu in Asia; and cholera in Haiti, among other health epidemics.

2 Layered spaces that have complex and deeper meanings, relationships, and dynamics than what is apparent on the surface.

3 'Strategic essentialisms [are] authoritative knowledge that is fine-grained and nuanced but exceeds its empiricism through theoretical generalization. Such forms of essentialism and dislocation, it is argued, are needed to dismantle the dualisms that have been maintained between global cities and mega-cities, between theory and fieldwork, and between models and applications' (Roy 2009: 822).

References

Brenner, N. (2013) *Implosions/Explosions: Towards a Study of Planetary Urbanization*. Berlin, Germany: Jovis Publishers.

De Certeau, M. (1984) *The Practice of Everyday Life*. Berkeley, CA: University of California Press.

Easterling, K. (2014) *Extrastatecraft: The Power of Infrastructure Space*. New York, NY: Verso Books.

Lefebvre, H. (1991[1974]) *The Production of Space*. Oxford: Blackwell.

Lefebvre, H. (2003) *The Urban Revolution*. Minneapolis, MN: Minnesota University Press.

Merrifield, A. (2013a) *The Politics of the Encounter: Urban Theory and Protest Under Planetary Urbanization*. Athens, GA: University of Georgia Press

Merrifield, A. (2013b) The urban question under planetary urbanization. *International Journal of Urban and Regional Research* 37 (3), May, 909–922.

Pieterse, E. and Simone, A. (2013) *Rogue Urbanism*. Johannesburg, South Africa: Jacana Media and ACC.

Roy, A. (2009) The 21st century metropolis: New geographies of theory. *Regional Studies* 43 (6), 819–830.

Roy, A. (2011) Slumdog cities: Rethinking subaltern urbanism. *International Journal of Urban and Regional Research* 35 (2), March, 223–238.

Roy, A. (2014) Urbanisms, worlding practices and the theory of planning. *Planning Theory Planning Theory* 10 (1), 6–15. First published date: February 11, 2011. doi: 10.1177/1473095210386065.

Simone, A. (2001) On the worlding of African cities. *African Studies Review* 44, 15–41.

Simone, A. (2004a) People as infrastructure: Intersecting fragments in Johannesburg. *Public Culture* 16, 407–429.

Simone, A. (2004b) *For the City Yet to Come: Changing Life in Four African Cities*. Durham, NC: Duke University Press.

Simone, A. (2010) *City Life from Jakarta to Dakar: Movements at the Crossroads*. New York, Routledge.

Simone, A. (2013) Cities of uncertainty: Jakarta, the urban majority, and inventive political technologies. *Theory, Culture and Society* 30 (7–8), 243–263.

Simone, A. and Rao, V. (2012) Securing the majority: Living through uncertainty in Jakarta. *International Journal of Urban and Regional Research* 36 (2), 315–335. December.

Yiftachel, O. (2009) Critical theory and 'gray space' mobilization of the colonized. *CITY* 13, 2–3, June – September. Available at: www.tandfonline.com/doi/full/10.1080/13604810902982227

Epilogue

Creating planning knowledge through dialogues between research and practice

Maja Karoline Rynning, Tanu Priya Uteng and Karen Lucas

How can lessons from this book contribute to create new knowledge for the praxis of city planning for cities in the Global South?

Through the lens of case studies and empirical examples, this book has shown how most of the contemporary theories and knowledge bases that are currently being applied to transport planning and practice in the Global South are directly or indirectly derived from western models. However, as several chapters have highlighted, the premises, requirements and constraints of mobility and accessibility in developing post-colonial cities are *fundamentally different* from those in the Global North. Some parallels can be drawn between the challenges and issues facing both global spheres, but nevertheless, as the authors have argued, solutions and policy strategies necessitate 'locally produced' knowledge and solutions – this remains crucial for both global spheres.

From the case studies, we can conclude that cities and their hinterlands in the developing world are a loosely packed collection of *homogenous spaces*, such as housing areas, markets, civic centres, business districts and so forth, intersecting with *contested spaces*, such as streets, gated areas, slum areas and transportation hubs. Furthermore, the very definition of 'urban' is in a state of constant change as multiple peri-urban areas are coalescing with the urban areas. Differentiated mobilities within these spaces have significant consequences for the divide between the 'have' and 'have-nots' population in the creation and use of spaces and infrastructures, along with access to activities within the urban realm. Given that space/place-making is infused with power and rights, and the battle between them, the position that mobilities occupy in this conflict plays a pivotal role within the narratives that we have included in this book.

In parallel to the more obvious differences, the book also confirms that there are certain similarities between the results and consequences of transport development in the Global North and South. For example, cycling is important for the inhabitants in Hanoi (Chapter 2), for economic as well as public health reasons, which strongly resonates with current foci and much research from the Global North (e.g. London, Paris, Copenhagen, Oslo). Similarly, when asking what sustainable mobility means for developing cities, especially with regards to accessibility and economic opportunity (Chapter 9), the responses parallel on-going discussions in developed cities, and among researchers in developed countries. These are included in current debates on accessibility over mobility, public space over congestion, and safety and wellbeing over speed (Banister, 2008).

Ultimately, the 'big question' for all cities in the Global North and South boils down to how to create socially inclusive and environmentally sustainable places, which makes its resources available to everyone, and where everyone's wellbeing can be equally protected within them. This is a large and complex question that politicians, developers, executives, planners and researchers worldwide need to ask themselves. Discussions regarding this can be found in research articles, in political discourses and in public debates on urban development in the Global North. As this book has shown us, these issues can, and are, also being raised in the context of developing and post-colonial cities in the Global South, but they remain under-explored and under-discussed. This highlights a potential to learn from one another and to explore these aspects together across the North/South binary. This is an absolute must to avoid replicating mistakes, for example, by adopting strategies that simply do not work, or are only effective in a specific given context or set of circumstances, and therefore cannot successfully be transferred elsewhere.

We have suggested in this book that a first step in the right direction would be to consider the mobility and accessibility challenges from the perspective of the *full-range of people* living, working and travelling in these cities; including not only formal residents but also the informal settlers and slum dwellers, who are currently largely highly marginalised and further excluded by urban/transport development processes. An important factor here is to bring to the fore differences among various segments of the population, especially income differences, and that these tend to be more extreme for cities in developing countries than for cities in developed countries. This indicates (and the cases from this book have shown) that such population groups are more sensitive to changes in the transport systems and urban structures than their counterparts in developed countries. The possible consequences, for example, concern:

- Those involved in the informal sector, and whose access to 'daily bread' can be jeopardised by faulty planning decisions (e.g. street vendors);
- Access to services, education, health services and markets, especially considering the gendered roles in the Global South;
- Expropriation, appropriation to resources and power-dynamics;
- Restricted margin and capacity to handle changes – where a high proportion of the inhabitants are vulnerable to changes.

Throughout the book, an overarching theme stands out: the importance of being aware of the local challenges and the local context. Solutions from one part of the world cannot simply be transposed to a different part of the world without local adjustments.

Urban development projects are wicked design problems

A second observation arising from the book is the stubborn and 'sticky' nature of urban development problems, such as the transport planning and mobilities issues that it identifies. It has been suggested by scholars outside the transport realm that these should be understood and addressed as *wicked problems*,[1] that is, ill-defined, complex, uncertain and unstable problems (Lawson, 1993; Rittel and Webber, 1973; Schön, 1983). In general, wicked problems have no beginning and no definitive end, and are never fully resolved. They have no right or wrong answer, nor an optimal solution (Lawson, 1993; Rittel and Webber, 1973; Schön, 1983). As a result, they can (probably) never be completely understood. Rittel and Webber (1973) write that every wicked problem can be considered to be a symptom of another problem: there is always another level of detail or point of view to be explored or considered.

As such, denoting urban transport planning as a *wicked problem* describes particularities for handling and resolving it, both in the developing and the developed cities.

> The kinds of problems planners deal with – societal problems – are inherently different from the problems that scientists and perhaps some of the classes of engineering deal with.
>
> Rittel and Webber, 1973

Rather, cities should be seen as complex networks of interactions and connectivities, built of interdependent elements and variables. Given the multiplicity of interactions among the different aspects of urban and transport-planning decisions, a change of one facet of the city will inevitably influence several others (Ascher, 1995; Carmona, 2010; Speck, 2013; Tennøy, 2012). One example is traffic-related casualties and deaths, which affect quite severely several sectors of the population in developing cities, and particularly those who rely on walking as their sole mode of transport. These impacts stem from various individual elements of the transport and land use system, such as inadequate or non-existent pedestrian infrastructures and lack of public transport services, which force people to walk long distances on dangerous roads, combined with overreached and badly managed driving environments. They can also be a result of service facilities and housing being located near heavily trafficked roads – potential symptoms of failed educational policies, failed urban development policies, and failed land use and transportation planning, to mention some. As a result, traffic-related casualties and deaths can be addressed, to a larger or smaller extent, through making suitable interventions in the delivery systems of all the foregoing.

Given that wicked problems are unique owing to the variations in the context, internal and external, a solution for one wicked problem cannot directly be applied to a different one. Knowledge of context is essential when trying to grasp and 'deal with' a wicked problem – more so, as every development action (and on a broader level, every project) leaves traces upon the city and its inhabitants that cannot be 'undone' (Rittel and Webber, 1973). Once more, the importance of *local* stands out with regards to urban development, and for producing planning knowledge.

Urban practitioners have a particular comprehension of the urban wickedness

Solving wicked urban development problems, such as those of daily mobilities, requires a particular skillset (Lawson and Dorst, 2009; Rittel and Webber, 1973), which urban practitioners are trained to have (i.e. architects, urban designers, landscape architects, urban planners, transport planners). These skills form a distinctive *savoir faire*,[2] characteristic to urban practitioners. For urban professionals such as urban planners, transport planners and urban designers, this means knowing how to observe and to comprehend cities and their functionings[3] in order to solve their problems. The urban practitioners' savoir faire includes technical knowledge, process knowledge, methodological knowledge and design knowledge (Cross, 1982; Darke, 1979; Lawson and Dorst, 2009; Kirkeby, 2012; Kirkeby, 2015; Schön, 1983; Skogheim, 2008; Tennøy, 2012). It is primarily based on their education and professional experience. Through his or her practice, the urban practitioner further develops a set of design principles, which Lawson defines as their intellectual luggage (Darke, 1979; Lawson, 1993).

Ideally, their savoir faire and reflective design principles impart urban practitioners with the third-eye – a particular approach to urban development. In a well-designed project, s/he will seek to go beyond the 'client's command or project mandate' in order to grasp the totality of the project in a holistic way, and, in part, to identify and comprehend the potential effects upon its urban context and the lives of urban inhabitants (Schön, 1983; Kirkeby, 2012). But often, such an ideal state of practice does not take place as a result of a categorical lack of knowledge of the overlapping dimensions. By exploring in depth the physical, economic, social, cultural and political context, the reflexive urban practitioner can gain an understanding of the local and urban context, to suggest locally adapted measures and solutions. It will enable her/him to identify issues that may not be explicitly covered in the project description (e.g. underlying social and cultural factors), but which are potentially important for the end result and its reception by neighbourhood actors, and the city as a whole. This is an example of how experience provides the practitioner with knowledge about particular kinds of wicked problems, or particular aspects of a wicked problem, which he or she uses when a similar problem is encountered (Lloyd and Scott, 1994).

This approach is what Rittel and Webber refer to as finding what the problem 'really is', knowing that urban development (vis-à-vis urban mobilities) problems are indeed societal problems. Together with the savoir faire and the design principles, this makes experience-based understandings of the urban professional particularly valuable for producing knowledge on urban development. In part, through their 'professional eye' (Skogheim, 2008)[4]: a particular way of observing and comprehending the urban built environment, developed through the professional practice, their experience-based knowledge comes from a different perspective and rationale than that of research. The two are complementary, and when combined can provide a rich and detailed image of a city, its functionings and how to address development targeted towards good living conditions for its inhabitants.

Where to next? A need to combine evidence-based and experience-based knowledge

The main objective of the book has been to create a knowledge-based starting point for more inclusive planning for sustainable mobilities in cities of the Global South, and for post-colonial cities worldwide. The case studies have provided a descriptive collage of the main challenges related to urban mobilities. This has helped us to establish a preliminary understanding of the nature, depth and complexity of the planning issues that are inherently involved. However, much more needs be done to develop and explicate this initial knowledge base within the academic and the policy world.

An important element of any further research should be to create effective knowledge feedback-loops between academics, policymakers and relevant stakeholders which can incorporate 'new mobilities' perspectives into mainstream transport and land use planning decisions in Global South cities. The following five key criteria can be made part of this feedback system (inspired by Lucas and Currie, 2012):

1. What is the nature of the mobilities taking place? – conceptualisations, definitions, theoretical perspectives;
2. What are the reasons for the various facets of mobilities? – market effects, public policies, funding structures, service provision, capacities and constraints of individuals;
3. Who is affected or at risk? – demographical breakdowns, distribution across different income groups, behavioural analyses;
4. Where is it physically happening? – geographies, spatial distributions, affected areas, settlement types;
5. How can it be addressed? – action pathways, strategies and timescales, tools, resources and capacities, institutional arrangements, delivery agencies, existing good practices.

In the beginning, feedback attempts will likely increase the overall complexity of the understanding of 'mobilities', but as long as their dynamics are dependent on one another, they can be combined to create a feedback communication system to inform future planning decisions. Future research could concentrate on expanding this feedback system by looking into different groups, places and processes that might affect daily mobilities. The scope and possibility of developing such a feedback system should be explored in different countries. The cases presented in this book facilitate such nuanced understandings.

Given that city planning in the Global South represents similar wicked problems as city planning in the Global North, solving these wicked problems calls for interdisciplinary strategies and approaches, as does knowledge production on *how* to achieve them in practice (Dubois et al., 2016; Rittel and Webber, 1973; Schön, 1983). Therefore, creating knowledge for urban development of cities in the Global South requires similar interdisciplinary research collaborations as are currently being attempted in the Global North. These include regular interactions between researchers and policymaker/planning practitioners across a range of cross-cutting policy sectors, including housing, social welfare, education, health care and other facilities and service-sector planning.

This brings to light an additional and highly relevant concern for furthering state-of-the-art research in the Global South – the paucity of suitable data at a fine level of granularity, and the urgent need for better data collection and analytical protocols. These concerns need careful formatting through the following steps: 1) consideration of the kinds of data currently being collected and their usefulness (and inclusiveness); 2) methods to gather data so that the conditions and activities of *all people* living and working in the city, and not only its formal residents, are captured; and 3) assuring a certain standardisation of the data (for comparisons and theoretical development). Along with the traditional methods, we have a host of ICT-enabled new methods available today to finally make a dent in the data paucity scenario. The challenge to combine these methods, to study the factors affecting behavioural changes taking place in the current times, needs immediate attention. Studying behavioural changes includes a wide variety of topics, seemingly removed from transport concerns per se but in reality, is strongly interlinked with the mobilities agenda (e.g. safety concerns, residential self-selection and spatial sorting mechanisms and its interactions with the labour markets).

A final observation from the book is a confirmation of the innate inertia of urban and transport-planning systems regarding decisions and interactions – particularly relevant are the institutional arrangements for transport planning and policy development in the context of the Global South. This inertia is at once a characteristic and a property of the decision-making and development processes, influencing both the way projects evolve and their end results. Additionally, the conditions and premises for a project often change along the way, further complicating predictions about the actual outcome of design proposals, and their reception by the local, urban contexts. The long-term effects of development projects are often only properly perceived five, 10, or even 15 years after their delivery – this is particularly true of transport infrastructure projects, which take a long time to reach full maturity. An often-overlooked connected issue is that what might appear to be a mobility solution for today, such as reducing traffic congestion through building new roads, could become the mobility problem of tomorrow, such as the severance of communities from activities as a direct consequence of that very same road, and hindering the use of sustainable mobility modes such as walking and cycling.

The inertia of urban development means that cities will deal with the consequences of poorly planned projects for years to come. However, new development strategies and approaches are being applied with an increased understanding of the importance of locality in the Global North. This can be seen on several levels. At the local scale, cities in the Global North are exploring what car-free cities might represent in terms of a more equitable city (e.g. Oslo, Paris). One can hope that such positive experiences can contribute to counterweight the increasing auto-mobility of most cities in the Global South, allowing them to maintain local particularities of their urban lifestyle, such as informal economy, combined with sustainable modes of mobility.

On a more global scale, urban mobility has gained an increased importance within urban development circles (e.g. the UN Sustainable Development Goals). As this epilogue has argued, making a jump from being a pure infrastructure provider to studying the impacts of the nexus between infrastructure and empowerment requires new approaches, especially interdisciplinary collaborations across fields and geographical scales. It is our hope that the experiences from this book can inspire others to explore new strategies and methods to incorporate 'local' and 'wickedness' perspectives into city and mobilities planning for a sustainable urban future.

Notes

1 The opposite of a wicked problem is often considered a *tame* problem (Rittel and Webber, 1973): clearly defined and has a proper beginning and an end, for instance, a mathematical problem. That is not to say that they are easier to solve or to comprehend, merely that they are more structured and concise.
2 *Savoir faire* can be translated as 'know how' or 'knowing by doing' (European Union, 2014).
3 The term *functionings* refers to how a city 'works', that is, the various mechanisms, systems and interactions that take place daily in a city and how these 'function'. See, for instance, Glaeser (2011) or Ascher (1995) for more in this line of description.
4 Although Skogheim initially applied the term to architects, the works of Lawson, Cross and Kirkeby, to mention some, have shown that it applies to other design professionals as well.

References

Ascher, F. (1995) *Métapolis ou L'avenir des villes*. Paris, France: Odile Jacob
Banister, D. (2008) The sustainable mobility paradigm. *Transport Policy* 15, 73–80.
Carmona, M. (2010) *Public Places, Urban Spaces: The Dimensions of Urban Design*. London: Routledge.
Cross, N. (1982) Designerly ways of knowing. *Design Studies* 3, 221–227.
Darke, J. (1979) The primary generator and the design process. *Design Studies* 1, 36–44.
Dubois, C., Cloutier, G., Rynning, M.K., Adolphe, L. and Bonhomme, M. (2016) City and building designers, and climate adaptation. *Buildings* 6, 28.
European Union (2014). COMMISSION REGULATION (EU) No 316/2014 of 21 March 2014 on the application of Article 101(3) of the Treaty on the Functioning of the European Union to categories of technology transfer agreements.
Glaeser, E. (2011) Cities, productivity, and quality of life. *Science* 333 (6042), 592–594.
Kirkeby, I.M. (2012) Om at skape arkitektfaglig viten. *Nordic Journal of Architecture Research* 24 (2), 70–90

Kirkeby, I.M. (2015) Accessible knowledge – knowledge on accessibility. *Journal of Civil Engineering and Architecture* 9 (5), 534–546.

Lawson, B. (1993) Parallel lines of thought. *Languages of Design* 1, 357–366.

Lawson, B. and Dorst, K. (2009) *Design Expertise*. Oxford: Elsevier, Architectural Press.

Lloyd, P. and Scott, P. (1994) Discovering the design problem. *Design Studies* 15, 125–140.

Lucas, K. and Currie, G. (2012) Developing socially inclusive transportation policy transferring the United Kingdom policy approach to the State of Victoria? *Transportation* 39 (1), 151–173.

Rittel, H.W.J. and Webber, M.M. (1973) Dilemmas in a general theory of planning. *Policy Sciences* 4, 155–169.

Schön, D.A. (1983) *The Reflective Practitioner How Professionals Think in Action*. New York, NY: Basic Books.

Skogheim, R. (2008). *Mellom Kunsten og Kundene Arkitekters Yrkessosialisering og Profesjonelle Praksis*. PhD-thesis. University of Oslo.

Speck, J. (2013) *Walkable City: How Downtown Can Save America, One Step at a Time*. New York, NY: North Point Press; Reprint edition.

Tennøy, A. (2012) *How and why planners make plans which, if implemented, cause growth in traffic volumes Explanations Related to the Expert Knowledge, the Planners, and the Plan-Making Processes*. PhD-thesis. Norwegian University of Life Sciences NMBU.

UN Habitat (2013 *Planning and Design for Sustainable Urban Mobility: Global Report on Human Settlements*. (2013) Abingdon, Oxon: Routledge

Index

Page numbers in italic indicate a figure and page numbers in bold indicate a table on the corresponding page.